ENERGY END USE

An Environmentally Sound Development Pathway

Asian Development Bank

Energy End Use presents the opinions of the authors, and not necessarily those of the Asian Development Bank or its officers.

$ refers to US dollars and Rs refers to Indian rupees except where otherwise noted.

ADB Coordinator
Ali M. Azimi

Authors
Gautam Dutt and, for Chapter 9, N. H. Ravindranath

Editor
Jill Gale de Villa

Design and Artwork
Indira Sandilya and the ADB Printing Unit

Typeset and Printed by
ADB Printing Unit

Published by
Asian Development Bank
PO Box 789, 1099 Manila, Philippines

ISBN 971-561-013-7

FOREWORD

Asia is in the midst of an electricity crisis resulting from booming economic growth that led to rapid growth in the use of energy. Virtually every country in the region is facing a potential capacity crunch. An affordable, reliable supply of energy is vital for continued economic expansion, yet the cost of new supply facilities is staggering.

If energy supply capacities are expanded without promoting more efficient use of energy, the amount of capital invested will eclipse the economic growth it seeks to sustain. The amount of investment needed for new electricity production in Asia is enormous. An estimated US$500 billion will be used for fourteen Asian countries (excluding Japan) for generation, transmission and distribution of power in this decade alone. China and India are expected to account 80 per cent of the Asian total.

The environmental and social problems associated with this expansion are also formidable. Conventional coal fired power plants are a major cause of acid rain and contribute to global warming. Nuclear power involves huge capital costs, unresolved waste-disposal problems and growing public opposition. Hydroelectric power projects have stoked controversy in Thailand, India and China because of resultant relocations of indigenous population and ecological damage.

Energy planners have provided subsidized electricity supplies, and have tried to build enough power plants to match the growth in demand. However, technological innovations in energy efficiency have changed the relative costs of power generation and electricity conservation measures so that conservation is often cheaper than building new power plants. Using more efficient lights, agricultural and industrial motors, and insulation can reduce significantly the supply needed.

The study was developed through a collaborative effort between the Energy and Industry Department and the Office of the Environment to illustrate potential energy savings in India through energy end use. India is experiencing phenomenal growth in demand for electricity. On average, demand has grown 10 per cent per annum, a rate which is likely to be maintained in the 1990s.

India seeks to add 27,000 MW of generating capacity by the year 2000, a formidable task considering the difficulty experienced in adding 2,000 MW in the last five years. Perhaps the simplest way out of this developmental challenge is to use electricity more efficiently. Improvements in electricity end-use efficiency offer a cost-effective and environmentally benign way of addressing this challenge.

The study was initiated by Ali M. Azimi, Office of the Environment, Asian Development Bank; Gautam Dutt collected and organized the required information, and wrote the text. N.H. Ravindranath assisted with Chapter 9. Jill Gale de Villa edited the manuscript.

The Bank hopes that information from the study will serve as a basis for stimulating awareness in Developing Member Countries and in the International Community.

KAZI F. JALAL
Chief, Office of the Environment
Asian Development Bank

ACKNOWLEDGEMENTS

The author gratefully acknowledges the help provided by the following individuals:

- Howard Geller and Steven Nadel of the American Council for an Energy Efficient Economy, Washington DC;
- Samuel Baldwin of the Office of Technology Assessment, Washington DC;
- Padmanabhan of the US Agency for International Development, Washington DC;
- Josephine Arpaillange, Douglas Barnes, and Willem Floor of the World Bank, Washington DC;
- Eric Larson of Princeton University, Princeton, NJ;
- Sam Baldwin, Office of Technology Assessment, US Congress, Washington DC;
- Thomas Johansson of Lund University, Sweden;
- Ashok Gadgil and Omar Masera of Lawrence Berkley Laboratories, California;
- Ali Azimi, V. Desai, Lars Hultin, Bindu Lohani, Kari Nyman and others of the Asian Development Bank, Manila;
- Mark Cherniack of the International Institute for Energy Conservation, Bangkok;
- N.H. Ravindranath, A.K.N. Reddy, H.S. Mukunda, and U. Shrinivasa of the Indian Institute of Science, Bangalore;
- Anil Date of the Indian Institute of Technology, Bombay;
- S. Anand, Rangan Bannerjee, G. Chandrasekar, Veena Joshi, Bhaskar Natarajan, Nalini Ranganathan, Mala Swaminathan, and others of the Tata Energy Research Institute, New Delhi;
- Kirk Smith of the East-West Centre;
- Krishna Prasad of the Eindhoven University of Technology;
- Robert Williams;
- Virendra Kothari of Innotem Services, New Delhi;
- Rakesh Mohan, Economic Advisor to the Minister of Industry, India; and

- for assistance with conducting the cooking and water-boiling tests, Mrs. Jaylaxmi, Mr. Mohan Nayak, and Mr. Nagaraja.
- The author received valuable information from the following organizations:
- Indian Pump Manufacturers Association, Bombay;
- All India Electric Motor Manufacturers Association, Bombay;
- Indian Electric and Electronic Manufacturers Association, Bombay;
- Industrial Development Bank of India, Bombay;
- Centre for the Monitoring of the Indian Economy, Bombay;
- Bureau of Indian Standards, New Delhi;
- Intermediate Technology Development Group, Rugby, England;
- National Productivity Council, New Delhi; and
- Energy Management Centre, New Delhi.

Text was edited by Jill Gale de Villa. Book design was conceived by Indira Sandilya. The ADB Printshop provided artwork and completed production.

TABLE OF CONTENTS

Appendixes

Boxes

Figures

Tables

ABBREVIATIONS AND ACRONYMS

ADB	Asian Development Bank
AL	annualized life-cycle cost
ATF	aviation transport fuel
CFL	compact fluorescent lamp
CFR	conventional fan regulator
CO	carbon monoxide
CO_2	carbon dioxide
CRF	capital recovery factor
CSE	cost of saved energy
EEC	European Economic Community
EFR	electronic fan regulator
EHW	electrically heated water
EMC	Energy Management Centre
GDP	gross domestic product
GJ	gigajoule
GNP	gross national product
GW	gigawatt
HID	high-intensity discharge
HP	horse power
HPS	high pressure sodium
ICICI	Industrial Credit and Investment Corporation of India
IDBI	Industrial Development Bank of India
IPS	irrigarion pump sets
IRR	internal rate of return
kWh	kilowatt-hour
LDC	less developed country
LPG	liquid petroleum gas
MJ	megajoule
MV	mercury vapor

NABARD	National Bank for Agriculture and Rural Development
OPEC	Organization of Petroleum Exporting Countries
PACER	Program for Acceleration of Commercial Energy
PVC	polyvinyl chloride
SPP	simple payback period
TERI	Tata Energy Research Institute
Toe	tons of oil equivalent
TSP	total suspended particles
TWh	terrawatt-hour
USA	United States of America

1

Introduction

Studies that project energy demand based on historical and cross-sectoral data have predicted large increases in future energy demand. Given the observed correlation between per capita energy use and per capita gross national product (GNP), both historically for individual countries and among countries, these studies concluded that a continuing increase in energy use was a prerequisite to progress in developing countries. Even the proponents of these energy strategies admitted that material and capital resources were likely to limit attaining high energy growth rates, implying that economic development was unattainable by most of the world's population. Pursuing a supply-side energy strategy will entail a high price in terms of investments and environmental costs such as acid rain, land disturbance, deforestation, nuclear proliferation, and global climate change. For most countries, which must import oil and coal, this strategy also means a continuing heavy burden of foreign debt that already threatens their economic future.

Against this backdrop, an alternative demand-side energy strategy has begun to gain momentum. The strategy is based on the actual uses for which energy is needed and the ways the needs could be most efficiently met. This strategy, directed towards sustainable development, questions equating energy use with GNP, and GNP with development (Goldemberg et al., 1988). The features of this energy strategy are as follows:

- satisfying basic human needs by directly allocating energy resources,
- creating fair economic comparison between all energy sources and end uses,
- promoting energy efficiency improvements,
- beginning a transition to renewable energy sources,
- generating new knowledge and technological advances, promoting national self-reliance, and
- ensuring compatibility of energy strategies to other global problems.

Using this framework, a study was conducted by the Asian Development Bank (ADB) using India as an example, to formulate energy strategies that are compatible with, and contribute to, the solution of the other major problems—including North-South disparities, the poverty of the majorities in the developing countries and of minorities in the industrialized countries, environmental degradation in both the industrialized and developing countries, the threat of global climate change, and global insecurity and risks of nuclear weapons proliferation.

The sustainable-world energy strategy is based on a detailed look at how energy is used rather than on the traditional preoccupation with energy supply and aggregated demand. Because of its emphasis on the uses to which energy is put, it is also known as an end-use-oriented energy strategy. An end-use-oriented approach considers the end uses of energy or the energy services performed, such as cooking, lighting, and passenger and goods transport. By considering the technical and economic details of present and alternative end-use devices—stoves, lamps, vehicles, etc.—one can identify improved ways of meeting future demand for energy services. These may be classified into three groups:

- more efficient end-use devices, e.g., more efficient motors, boilers, cars, and cooking stoves;
- synergies, e.g., technologies that permit simultaneous servicing of several end uses, such as the cogeneration of heat and power; and
- new supply options, especially those using renewable and decentralized resources, e.g., technologies that are not considered in conventional, centralized energy planning.

The end-use-oriented approach considers these alternatives as well as conventional technologies to arrive at an optimal mix of centralized and decentralized, renewable and nonrenewable supply options, taking advantage of cost-effective energy efficiency measures and synergies of energy conversion. The end-use-oriented approach leads to the identification of an energy future that is much less capital and resource-intensive (and less expensive overall) and more environmentally benign.

In this report, we present a preliminary, global analysis of the implications of an end-use-oriented strategy for sustainable development for an Asian developing country. We illustrate our analysis with the case of India. India's large oil import bill adversely contributes to an acute balance-of-payments problem, and energy (particularly electricity generation) projects make up the lion's share of increasing external debt. Traditional biomass fuels comprise almost half of the energy input to the Indian economy, and, despite rapid growth (about

10 per cent a year) in the installed capacity of electric power plants, most of the country suffers from mild to serious shortages of electric power.

This report emphasizes the end uses of electricity, both because of the problems noted above, and because this is the principal area of interaction between a developing country's energy sector and the ADB, through loans that have traditionally gone towards the construction of large power plants and transmission and distribution systems. We also look at the transportation sector, which accounts for the bulk of the petroleum used in most countries (including India), and at cooking, which is one of the most basic energy needs and accounts for most of the biomass use in most developing countries.

In Chapters 2 to 4, we look at several important end uses of electricity—lighting, residential appliances, pumping water, industrial motors—considering current use as well as technological and policy alternatives for attaining higher efficiency.

Lighting (Chapter 2) accounts for about 17 per cent of total electricity use in India (about 29 TWh out of 170 TWh in 1989/90), principally in incandescent (10 per cent of total) and fluorescent lamps (6 per cent). Cost-effective measures include replacing of incandescent lamps that are used more than two hours a day by compact fluorescent lamps (saving about 75 per cent), and replacing those used two or less hours per day with efficient incandescent lamps (saving 10 per cent). The cumulative savings potential is 9.5 TWh/yr. A number of energy-saving opportunities are possible for fluorescent tube installations, from a combination of better lamps, electronic ballasts, reflective luminaires, and better controls. The total savings potential is at least 6 TWh/yr. A potential additional 1 TWh/yr may be saved in street and other outdoor lighting. Thus, the total savings potential in lighting is 16.5 TWh/yr—57 per cent of the present level of electricity used for lighting. This large potential savings can be realized relatively quickly. Appendix I discusses the economic ground rules for a rational comparison among alternative end-use efficiency, energy supply, and synergy alternatives.

Electricity use has been growing faster in the residential than in other sectors, and accounts for about 13 per cent of total electricity sales in India. Some of this is for lighting, and the balance is for household appliances (Chapter 3). Reliable data on the relative contribution of the different appliances to total electricity use are not available. Computations from appliance sales data and empirical values of typical unit consumption levels show that fans and refrigerators, the largest users in the residential sector, consume about as much electricity as lighting. In both cases, savings are best attained through energy efficiency introduced into new units sold. As efficient fans and refrigerators enter the market, savings increase cumulatively, eventually reaching 13.5 TWh/yr for fans and 10 TWh/yr for refrigerators, assuming sales remain steady at present values of 5 million fans and 600,000 refrigerators a year. Although the relatively few electric resistance water heaters, air conditioners, and washing machines extant do not currently demand a large part of residential electricity, more households are expected to acquire these appliances in the future. Thus, policy steps must be taken now to ensure that new air conditioners and washing machines are energy-efficient and that electric water heating is discouraged in favor of solar, heat pump, natural gas, or producer gas-fired water heaters.

Agricultural pumping (Chapter 4) is a major end user of electricity in India, accounting for about 32 TWh in 1989/90. Extensive experience has been gained in retrofitting pumps to reduce electricity use. The use of frictionless foot-valves and smooth PVC piping reduces average electricity use by 29 per cent. A complete replacement of existing motors, pumps, and piping with an energy-efficient system reduces electricity consumption by 47 per cent. All measures are highly cost-effective from a social perspective; given rural tariff structures, however, private incentive for energy savings is insufficient. We estimated that a potential of at least 11 TWh/yr may be saved in agricultural pumping.

Requiring new electric pumps to meet existing Indian standards for energy efficiency, limiting agricultural loans for new pumps to energy-efficient systems, and removing incentives that favor pumping with electricity will help reduce future load growth in electric pumps. Hastening the retrofitting of existing pumps to make them more efficient will require an extensive training program that must be actively supported.

Pumping for water supply and sewage uses 3 TWh/yr. Studies are needed to determine the best ways to reduce energy use in these end uses. Industrial pumps also should be studied for their energy savings potential.

About 56 per cent of all electricity in India is used in industry. Of this, over 72 per cent is used by motors (Chapter 5). Industrial motors consume over 40 per cent of all electricity use in India (75 out of 170 TWh in 1989/90). In 11 electricity-intensive industries whose motors have been surveyed, about 75 per cent of the motors are of the squirrel-cage variety. Recently, Indian standards have been issued defining the performance of energy-efficient motors in this category. International standards call for higher efficiency levels. If all new motors sold in India met the Indian or international standards for efficiency, corresponding savings would be 0.25 or 0.58 TWh/yr, respectively.

Energy efficiency of the motor is only one area of improvement. Through improved maintenance, electrical tuneup, and better rewinding techniques, an estimated 16 per cent of the electricity used in existing motors can be saved—a total of 12 TWh/yr. Additional savings are possible through improved motor controls, drive trains, and other elements of the motor drive system. As we noted above, a 47 per cent reduction was possible and very cost-effective in irrigation pumping systems. A study estimated an electricity savings potential of 28-60 per cent in US motor drive systems. Since the existing condition of motors is generally worse in India, the savings potential is likely to be near the upper end of the range.

To promote electricity efficiency in new motors, the sale of standard motors should be phased out, progressively introducing higher efficiency units. Energy efficiency improvements in existing motor drive systems (numbering well over a million in industry alone) will require training and equipping many professionals. This could be done by strengthening an ongoing effort at training energy auditors for industry. Better information on the current energy consumption pattern of existing motor drive systems and documentation of successes and failures of past conservation projects are essential to the success of a large-scale program to save energy in motors.

Table 1.1 presents an approximate breakdown of electricity by end uses based on available information. The total 1989/90 consumption may be compared to the sales (170 TWh) plus

Table 1.1
Approximate Breakdown of Electricity by End Users in India, Showing Consumption and Savings Potential

End use	Consumption TWh in 1989/90	Savings potential[a] TWh/yr, in 5 to 10 yrs
Lighting	29	16.5
Residential (excluding lighting)	12	2.9
Agricultural pumping	32	11
Public water and sewage pumping	3	?
Industrial motors	75	31.4
Transport (trains, trams)	4	?
Commercial (excluding lighting)	7	1.0
Industrial (excluding lighting and motors)	21	?
Total use	**183**	**62.8**

a Savings potential in 10 years exceeds current consumption, assuming continued rapid growth in the demand for household appliances. This figure includes savings from ventilating fan improvement in all sectors.

self-generated electricity (15 TWh), totalling 185 TWh. The closeness of the agreement between supply and demand is fortuitous since electricity theft, if included, would increase supply. On the other hand, some end-use consumption has been estimated from equipment sales data, and there tends to be underreporting of sales, for tax purposes. Nevertheless, Table 1.1 indicates the relative importance of the various end uses of electricity.

In Chapter 6 we present the results of an analysis Reddy, Balachandra and D'Sa (1990) conducted for the Indian state of Karnataka, comparing the costs of electricity conservation, and various centralized and decentralized electricity generation alternatives. Although no shadow environmental costs were included, and these are likely to be less for electricity efficiency improvement and some decentralized renewable generation technologies, Reddy showed that large conventional nuclear, coal, and hydroelectric generators turn out to be the most expensive ways of providing or saving a unit of electricity.

The relative costs of saving or generating electricity by different technologies can be used to develop a strategy for providing a given level of energy services at least cost. Such least-cost electricity planning has become the norm in electric utility planning in the US. Chapter 7 presents Reddy's least-cost electricity plan for Karnataka. Rather than merely projecting future demand for electricity services, this analysis takes a development focus and seeks to satisfy the needs of the poor as well. The capital requirements for this plan are only 38 per cent that of the conventional approach ($0.618 billion per year, instead of $1.648).

Since only four electricity conservation measures were considered in the Karnataka study, a more detailed study is likely to lead to further cost reductions. For India and other countries, the implications of the Karnataka study and our earlier analysis of other end uses are presented in Chapter 7. For the end uses and conservation measures found to be cost-effective, the electricity savings potential is shown in Table 1.1. The total savings potential is at least 63 TWh, equivalent to an electricity generation capacity of 14.4 gigawatts (GW), operating at 60 per cent capacity factor (plant load factor) and assuming transmission and distribution losses of 20 per cent. In 1989, India's electric utilities had an installed capacity of 60 GW. As shown in Chapters 6 and 7, the savings can be accomplished at costs substantially below that of providing an equivalent amount of electricity from new power plants. These findings should be used to shift the emphasis from the traditional supply-side orientation towards a least-cost strategy.

2

Energy-Efficient Lighting

Lighting offers some of the most spectacular opportunities for electricity conservation. Lighting consumes 8-17 per cent of total electricity produced in industrialized countries (McGowan, 1989). Its share in developing countries can be larger. Energy use can be reduced by as much as 75 per cent without reducing lighting levels. A reduction of this magnitude is difficult to achieve in most other end uses. Moreover, some uses of lighting coincide with peak electricity demand, adding considerably to the generation cost or contributing to load shedding ("brownouts"). Finally, lamps have a relatively short life compared to other capital equipment. The high turnover implies that savings can be quickly realized through an effective implementation program.

While many lighting applications offer the potential for large savings, here we consider only a few important areas which together account for the bulk of lighting energy use.

Electric lamps are generally classified into three categories—incandescent, fluorescent, and high-intensity discharges (HID). Incandescent lamps are widely used, particularly in the residential sector. Fluorescent lamps are mostly used in offices, although some are used in houses and industry. The HID lamps typically provide higher illumination levels and are used in industry, and in street and other outdoor lighting applications. The three main types of HIDs are mercury vapor, high pressure sodium, and metal halide. The progression in the efficiency of each lamp type is shown in Figure 2.1. Below we discuss energy savings opportunities where incandescent and fluorescent lamps are currently used, and for street lighting.

Incandescent Lamps

The age of electricity started when Edison invented the incandescent lamp. Despite many innovations to improve its efficiency, the incandescent lamp remains the least efficient of all electric lamp sources. Since the introduction of fluorescent lamps in the 1930s, many offices and industries have switched to these lamps. Nevertheless, incandescent lamps continue to be used extensively, particularly in the residential sector. Fluorescent tube lamps are found infrequently in residences. They have made a relatively small penetration due

Figure 2.1

Approximate efficacies of various light sources plotted as a function of time, showing the evolution within each technology as well as the progress from one technology to another.

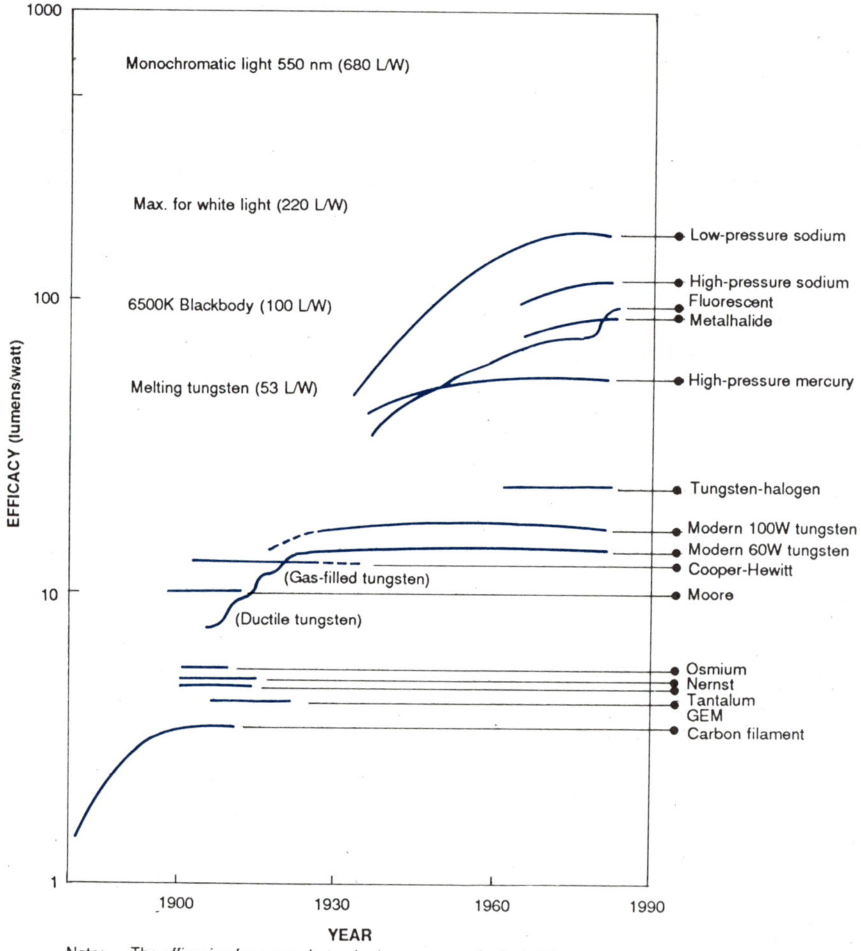

Note: The efficacies for several standard sources are indicated for comparison.

Source: J. M. Anderson and J. S. Saby, 1979

to a combination of high capital cost, more complicated installation, and poor light quality.

The extent of incandescent lamp use and energy consumption can be estimated from extensive surveys of users in all sectors. If precise information by sector is not required, an overall estimate of the total electricity consumed by incandescent lamps can be calculated from lamp sales data. For instance, about 250 million incandescent lamps are produced in India per year (Economic Survey 1989-90, 1990). The weighted average wattage is about 65 (Gadgil and Jannuzzi, 1989). Since the rated lamp life is 1,000 hours, the average lamp will consume 65,000 watt-hours or 65 kWh in its lifetime. In the steady state of 250 million lamps sold every year, the lamps sold in any given year replace the ones that burnt out that year, and the annual energy use by these lamps is equivalent to 1,000 hours each, per year. The total electricity used by incandescent lamps is 65 x 250 million kWh or 16.2 TWh. In reality, part of the sales are for new installations and any shifts in the patterns of sales among different kinds of lamps will affect this estimate. However, the error introduced by assuming the rated life is probably higher. Developing countries often have large voltage fluctuations, and over-voltages drastically reduce bulb life (van der Plas, 1988). Thus, the actual consumption by incandescent bulbs is somewhat lower than our estimate.

The total electricity sales in India was 146 TWh in 1987/88 (Statistical Pocket Book -- India, 1989), and about 170 TWh in 1989/90 so that incandescent lighting represents almost 10 per cent of total sales. (The 170 TWh figure is the author's estimate, and does not include self-generated electricity, another 15 TWh, mostly used in the industrial sector.) Gadgil and Jannuzzi (1989) obtained a similar fraction using different estimation methods.

Electricity use in incandescent lamps is best reduced by replacing high-usage lamps with compact fluorescents, and low-usage lamps with efficient incandescent bulbs. Taking advantage of new materials, lighting manufacturers developed fluorescent lamps that are only slightly larger than incandescent lamps. Initially introduced in the early 1980s, a range of these compact fluorescent lamps (CFLs) are now offered by several major manufacturers. Many produce a warm white light that is indistinguishable from incandescent bulbs. Some can be screwed into incandescent lamp sockets. They have long life (rated at 9,000 hours or more) with the added convenience that frequent replacement is not necessary. The CFLs consume only a fourth or even less electricity to produce the same amount of light as incandescents. For instance, a CFL that consumes only 16 watts has the same rated light output as a 75-watt incandescent. The CFLs have light outputs equivalent to all the standard incandescents (40, 60, 75, and 100 watts). The CFLs use more exotic materials than ordinary fluorescent tubes and cost considerably more. While an incandescent typically costs $0.40 to $0.75, a CFL retails for $14 to $18. The energy cost of an incandescent lamp far exceeds its purchase price for lamps which are on several hours a day. Thus, the CFL is cost-effective at electricity prices in many industrialized countries (see Table 2.1).

Table 2.1
Economic Comparison of Regular and Energy-Saving Incandescent Lamps

Lamp Type	Lamp Price (US$)	Lamp Life (hrs)	Annual Life 5 hr/day	Cycle Cost (US$) 1 hr/day
Regular incandescent	0.39	1,000	13.10	2.64
Energy-saving incandescent	0.54	1,000	12.08	2.44
Compact fluorescent	17.00	8,000	7.84	2.75

- Power: regular incandescent - 75W; energy-saving incandescent - 67W; compact fluorescent - 16W.
- Lamp cost and life data from Eletrobras, 1989, "Revisao das metas de conservacao de energia eletrica", Depto de Utilizacao de Energia, Rio de Janeiro, Brazil, cited in Geller, 1990a. With manufacture of CFLs in developing countries prices may drop somewhat. The rated life of compact fluorescent has been reported to be between 6,000 and 10,000 hours (Reddy, Balachandra and D'Sa, 1990; Gadgil and Jannuzzi, 1989).
- A real discount rate of 12%, and constant electricity prices of US$0.09 per kWh have been assumed. Because of subsidies, residential consumers in India, Pakistan and other developing countries pay considerably less. Long-run marginal costs of power supplied to the end user are likely to be at least 25% higher. See Chapter 7.

Although sales of CFLs have been increasing rapidly in industrialized countries, these lamps are virtually unknown in developing countries. There are several reasons. First except for Mexico, where Philips started manufacturing CFLs in 1989, no CFLs are made in developing countries. This reduces availability, and import tariffs common to many countries add considerably to cost. Electricity rates are subsidized for certain consumer categories in many developing countries, where domestic consumers pay less than the average cost of supplying electricity and far less than the long-run marginal cost. Furthermore, capital scarcity among potential users and other factors lead to high implicit discount rates for investment in energy efficiency. All these factors contribute to a virtual absence of CFL sales in developing countries. Thus, the introduction of CFLs presents a challenge to rational energy planning.

Recently-developed, more efficient incandescent lamps use about 10 per cent less electricity while providing the same level of illumination. They cost slightly more—in Brazil, the energy-saving lamp costs $0.54 vs. $0.39 for the standard incandescent. For lamps that are used only one or two hours a day, the investment in CFLs is not cost-energy effective. Table 2.1 presents an economic comparison among standard incandescents and alternatives for lamps used five hours and one hour a day. For the higher usage, the CFL has the lowest annualized life-cycle cost. For low usage, the energy-saving incandescent is the most cost-effective. Thus a coherent plan for increasing energy efficiency in lighting would be to require that all incandescents sold are energy-efficient and promote CFLs for high-use lighting.

It should be emphasized that in situations of market imperfection where (for whatever reason) people are unwilling to invest in a cost-effective energy efficiency measure, one manufacturer cannot impulse a market shift to energy efficiency. Consumers often favor the lower first cost alternative despite higher overall cost, and the manufacturer promoting energy efficiency is likely to lose market share. An example of this phenomenon occurred recently in Brazil, when Philips de Brasil promoted energy-efficient incandescent lamps. After promoting these lamps for a couple of years, and selling much more efficient lamps than ordinary ones, they discontinued the promotion as they were losing the market share of incandescent lamps. The relative shares of energy-efficient and regular incandescent lamps are shown in Table 2.2.

Table 2.2
Sales of Standard and Energy-saving Incandescent Lamps of One Lamp Manufacturer in Brazil

Production Year	P er Cent Energy	Per Cent Standard	Average Wattage
1987	64.1	35.9	62.9
1988	76.0	34.0	61.1
1989	23.2	76.8	64.7

Source: Geller, 1990b.

Energy savings in incandescent lamp installations can also be realized through improved controls. Timers that turn off stairwell lamps after a set interval have been used for decades in many countries. This system may be viewed as somewhat of a nuisance and therefore not suitable for all situations. An alternative is an occupancy sensor that turns on the light when a person approaches and turns it off shortly after the person's absence is sensed. This type of control has been tested in a Brazilian hotel with enormous energy savings and very rapid payback, despite the expense of a sophisticated sensor and control system.

Savings potential. Most of the electricity is consumed in frequently-used lamps. If we assume that 75 per cent of the electricity is consumed in lamps used more than two hours per day, and that a savings of 75 per cent is realized when a CFL replaces an incandescent lamp, the total savings potential is:

$$0.75 \times 0.75 \times 16.2 \text{ TWh/yr} = 9.1 \text{ TWh/yr.}$$

By considering efficient incandescents for the remaining lamps (which account for the remaining 25 per cent of electricity use) and 10 per cent savings per point, total savings would be:

$$0.1 \times 0.25 \times 16.2 \text{ TWh/yr} = 0.4 \text{ TWh/yr.}$$

Thus, the total savings potential in incandescent lamps is 9.5 TWh/yr.

Fluorescent Lamps

Fluorescent lamps make up the bulk of commercial, public, and industrial lighting. In India, they are also used in street lighting and in the residential sector. The lighting system usually consists of a 40-watt lamp, a ballast for controlling the lamp, a starter, and a luminaire to hold the lamp.

Energy use in fluorescent lamps can be estimated from lamp sales and lamp life data. In 1986, 58 million fluorescent tube lamps were sold (production data for 1986 from Electric Lamp and Component Manufacturers Association, cited by Tata Energy Research Institute [TERI, 1990a]). We estimate 60 million lamps were sold in 1989/90 to replace burnt-out lamps, rather than additions to the stock. The rated lamp life is 5,000 hours, although anecdotal evidence suggests that actual life is considerably less. In the absence of empirical data we assume an average life of 3,000 hours, an average power input of 40 watts, and a ballast power consumption of 16 watts. The electricity used in fluorescent lamps (including the ballast) is thus given by:

$$(40 + 16) \times 3{,}000 \times 60 / 1 \text{ billion kWh} = 10.1 \text{ TWh}.$$

Thus, fluorescent lamps in all sectors consume almost 6 per cent of 1989/90 electricity sales of 170 TWh. (Incandescent and fluorescent lamps together are 16 per cent of the total.)

Although fluorescent lamps are quite efficient, considerable energy can be saved readily through more efficient lamps, energy-saving ballasts, improved luminaires, and lighting control systems. We have not considered long-term savings possible through architectural modifications that permit better utilization of daylight. Nor have we considered savings possible through reduced lighting levels. Where fluorescent lamps are located in air-conditioned spaces, reducing lamp power consumption reduces air conditioning loads, energy use, and could reduce peak electricity demand. These beneficial side effects are not considered in the following analysis. Instead we focus on the direct savings potential from measures that are easy to implement.

Energy-saving fluorescent tubes produce the same amount of light but consume about 10 per cent less energy than incandescents. Thus, a 36-watt lamp can replace a 40-watt one without any reduction in illumination. The 26-mm diameter tube is more efficient than the usual 38, and in India both produce about 2,440 lumens of light (for the "cool daylight" version). An 18-watt and a 58-watt tube produce the same light as a 20-watt and a 65-watt standard tube, respectively. Most lamps now installed in the US are the energy-saving kind. In India, the energy-efficient lamp costs the same as the standard one in early December 1990, 36-watt and 40-watt lamps were each retailing for Rs 29 or $1.60 in Delhi. The author got the impression through commercial inquiries that the salesmen believe that the 36-watt lamp produces less light, and is in some sense a poor substitute. (Perhaps anticipating this problem, the numerical value appearing on the lamp label is the wattage of the standard lamp it replaces, with other identifying letters that can be related to lamp power by consulting the lamp data sheet.) In some countries, 32- and 16-watt lamps are also available to replace 40- and 20-watt lamps, at a small penalty (about 7 per cent) in light output. They cost about 20 per cent more than the standard lamps (Geller, 1990a).

Ballasts. A ballast (sometimes called a choke) is needed to operate a fluorescent lamp. Some ballasts can be used with a pair of lamps. A standard iron core ballast costs Rs 30 ($1.65) and can consume anywhere from 12 to 25 watts, although most consume in the range of 15 to 18 watts (B. Natarajan, TERI, December 1990 pers. com.). One drawback of these ballasts is that they have low power factor (0.5), which increases transmission and distribution losses and increases the size of the conductors needed to supply a given electrical power. The power factor can be corrected using a capacitor, which costs Rs 20 (about $1). These are called "electromagnetic" ballasts, and although energy-saving electromagnetic ballasts exist, substantial savings require the use of electronic (solid state) ballasts. An electronic ballast consumes only 2 to 3 watts and costs Rs 140 to 250 ($7.80 to 13.90). Many of the electronic ballasts sold in India produce line frequency harmonics which can affect the operation of other electronic equipment (A.J. Gadgil, Lawrence Berkeley Laboratories, Dec. 1990, pers. com.). Electronic ballasts sold in industrialized countries must meet stringent standards. Before promoting them in developing countries, quality comparable to that in industrialized countries must be maintained in order to avoid undesirable side effects.

The combination of an efficient lamp and a good electronic ballast can reduce the power consumption of a 40-watt lamp system from 56 watts to about 38 watts with no loss of light. The economics of various combinations of regular and energy-efficient lamp, and regular electromagnetic or electronic ballasts are shown in Table 2.3 for two sets of the assumptions regarding lamp and ballast life. The first assumes these to be 3,000 and 10,000 hours, respectively, corresponding to what is believed to be the case for conventional lamps and ballasts. Although Indian manufacturers of electronic ballasts claim extended lamp life, published literature offers no data on lamp or ballast life. One manufacturer claims 10,000-hour lamp life with their electronic ballast as opposed to 5,000 with a conventional ballast. The results in Table 2.3 indicate that energy-saving lamps are cheaper to own and operate than regular lamps (this is obvious since the energy-efficient lamp costs no more than the regular one). However, any combination with electronic ballast is more expensive, if we assume no improvement in lamp or ballast life. Fluorescent lamps in the US are rated for 12,000 hours and a typical ballast life is 20,000 hours and corresponds to the second set of life assumptions in Table 2.3. Clearly, all the annualized life-cycle costs drop when these extended lives are assumed. The relative economics also change, and the energy-saving lamp and electronic ballast combination is the most cost-effective.

Table 2.3
Economic Comparison of Regular and Energy-Saving Fluorescent Lamps in Conjunction with Regular and Electronic Ballasts.

Lamp type/ Ballast Type[a]	Lamp Life (hrs)	Ballast Life (hrs)	Annual Life Cycle Cost (US$) 10 hr/day
Typical India			
Regular/regular	3,000	10,000	15.06
Energy-saving/regular	3,000	10,000	14.12
Regular/electronic	3,000	10,000	16.16
Energy-saving/electronic	3,000	10,000	15.23
Developed countries			
Regular/regular	12,000	20,000	13.80
Energy-saving/regular	12,000	20,000	12.86
Regular/electronic	12,000	20,000	13.18
Energy-saving/electronic	12,000	20,000	12.25

a The regular 40-watt lamp and the energy-efficient 36-watt lamp each cost US$1.65. The regular 16-watt ballast costs US$1; the 2-watt electronic ballast costs $13.90.

Note, however, that the annualized life-cycle costs are not very different from one another for either set of life assumptions. This means that the user does not make a substantial savings from the investment in energy efficiency. This is not uncommon in energy conservation investments, where annual life-cycle costs for a wide range of energy efficiencies are quite comparable. In this case, it is unlikely that users will invest in efficient ballasts, and even the "free" investment in energy-efficient lamps will only reduce overall costs by 6 to 7 per cent.

Realizing the 30 per cent savings from the combination of lamps and ballasts will therefore require some form of market intervention. The most common such intervention is the setting of minimum efficiency standards for lamps and ballasts. This route has been adopted by the US. An alternative, adopted by Brazil, consists of agreements between the government and the manufacturers, setting a goal for the phased reduction of the market share of the less-efficient products and corresponding increase of the more efficient ones. An agreement is preferable to a standard because it avoids an adversarial relationship between industry and government. A higher level of energy efficiency should be in the interest of manufacturers because it increases the value of their market, provided that:

- they have the resources and time to switch over their production facilities, and
- all manufacturers are subject to the same standards.

We have considered some opportunities for energy conservation through more efficient fluorescent lamps and ballasts. Additional savings in lamp efficiency are possible through advanced phosphors, including two-photon phosphors, isotopic enhancement of the mercury in the lamp, use of magnetic fields to reduce losses within the tubes, and electrodeless operation. Through these measures, lamp efficacy (light output per unit of electrical power input) increases from 61 lumens/watt (corresponding to a 40-watt lamp producing 2,440 lumens) to about 230 lumens/watt (Berman et al., 1989). In this report we have only considered the first step, a well-commercialized technology that brings lamp efficacy to 68 lumens/watt (a 36-watt lamp producing 2,450 lumens). Lamps commercially available in industrialized countries and some less-developed countries provide 100 lumens/watt (Berman et al., 1989). These and other energy-saving options will present developing countries with the potential for further and drastic reductions in the electric power input needed to generate a lumen. The energy used by the ballast cannot drop much below the levels already attained by today's electronic ballasts. Advanced ballasts will provide additional features, e.g., resist voltage fluctuations, permit lamp dimming, and control features described below. At least one electronic ballast manufacturer in India claims an operating range of 70 to 300 volts with either AC or DC supply (Hertz, 1990).

Luminaires and reflectors. Improved lamps and ballasts ensure that a given amount of light is produced at reduced power consumption. Not all the light generated reaches the surfaces where it is needed. A luminaire holds the lamp, and depending on its geometry and optical characteristics, can direct more or less light to where it is needed.

The simplest luminaire for fluorescent lamps consists of a metal strip which holds the lamp, ballast, and starter, and provides electrical connection. It has neither a reflector behind the lamp nor a diffuser in front to direct the light where it might be needed. Reflectors made of white painted enamel, polished and anodized aluminum, or mirrored glass can be placed behind the lamp in a fluorescent fixture to improve luminaire efficiency. The performance of different luminaires for two-lamp fluorescent fixtures was tested at the University of Sao Paulo. The results are presented in Table 2.4. Going from no reflector to a mirrored glass reflector doubles the useful efficacy of a fixture with two, 40-watt lamps. Reduced electrical input can be realized by using 32-watt tubes and an electronic ballast. In total, the useful efficacy is tripled. Based on the actual experience of one Brazilian bank chain, a four-lamp 40-watt fixture with two ballasts was replaced by a two-lamp 32-watt reflecting fixture with a single electronic ballast, resulting in energy savings of $0.019/kWh (Geller, 1990a).

Table 2.4
Comparison of the Performance of Different Fluorescent Fixtures

Lamp	Ballast	Type of Luminaire[a]	Rated Power (W)	Useful Light (lum)	Useful Light (lum)	Efficacy (lum/W)
2 x 40	standard	A	96	5,400	1,130	11.7
2 x 40	standard	B	96	5,400	1,710	17.8
2 x 40	standard	C	96	5,400	2,180	22.7
2 x 32	standard	C	77	5,000	2,100	27.3
2 x 32	electronic	C	66	5,000	2,100	35.0

a Type of luminaire: A - standard with completely exposed lamps; B - polished and anodized aluminum reflector; C - mirrored glass reflector.

Source: Moreira, 1988.

Metal-reflecting luminaires are marketed in India by a number of manufacturers at retail prices ranging from Rs 720 ($40) for luminaires with simple geometry to Rs 1,400 ($78) for a "bat wing" shape that improves the direction of reflected light. These prices are higher than those in Brazil ($15 to $45). The cost-effectiveness in India is thus lower than in Brazil. No documented studies of cost-effectiveness are available from India at present.

Reflective luminaires provide more light from the same number of lamps. Thus significant savings can only be realized if a four-lamp fixture can be replaced by a two-lamp one, as in the Brazilian case, or a two-lamp fixture with a single-lamp one. Four-lamp fixtures are relatively rare in India; most are single lamp, and some are two-lamp ones. Replacing these with a reflecting luminaire is not possible at present since reflecting single-lamp luminaires appear not to be available.

Most fluorescent installations in India consist of a single-tube lamp on a metal strip mounted on a wall (rather than suspended from the ceiling, as in two- and four-lamp fixtures). No reflecting fixture is available for wall-mounted single-lamp luminaires, and it would be difficult to guess the improvement of useful efficacy, since the efficacy of the present configuration depends on the color and reflectivity of the ceiling. One possibility for saving energy in this configuration would be to replace a 40-watt lamp fixture with a 20-watt one, if a reflective luminaire could be designed to effectively double useful efficacy. Additional electrical savings could be realized using 18-watt tubes and electronic ballasts. The technical and economic feasibility of energy savings through reflecting luminaires in one- and two-lamp situations needs to be determined through extensive field trials. One difficulty stems from higher levels of dust in India and other countries. Dust on the luminaire greatly reduces the reflectance and useful efficacy, and time between cleanings will need to be determined empirically. Additional savings will depend on the climate, the hours that the building is illuminated and air-conditioned, and the coefficient of performance of the air-conditioning system.

Controls. Improved lighting controls present another opportunity for considerable energy savings. Since users are often not responsible for paying the electric bills in commercial buildings, they do not bother to turn off unneeded lights. Electronics can make up for some human indifference. Some luminaires are equipped with sensors that measure the availability of daylight or the presence of human activity. The sensors turn off lights when there is sufficient daylight or when no one is in the room. Advanced electronic ballasts that allow fluorescent lamps to be dimmed vary light output inversely with the level of daylight available. In one demonstration of the potential energy savings to be realized through improved controls, Lawrence Berkeley Laboratories measured a reduction of 52 per cent in the energy used to illuminate the 58th floor of New York's World Trade Center (Berman et. al., 1989).

Savings potential. We estimated that fluorescent lamps account for about 10 TWh of India's annual electricity use. Since 36-watt lamps have made little penetration into the market so far, a 10 per cent savings is immediately possible by replacing 40-watt tubes with 36-watt ones. If electronic ballasts are found satisfactory (lifetime, harmonic generation, cost, etc.), then the total savings from lamp and ballast replacement would be about 30 per cent. Since satisfactory electronic ballasts are manufactured in other countries, and at least some ballasts manufactured in India are satisfactory (B. Natarajan, TERI, Dec. 1990, pers. com.), this should not be too difficult to realize over the next few years. Consumers would save more than energy as such ballasts will increase the life of the tubes as well. With advanced phosphors and reflecting luminaires, overall savings can increase to about 50 per cent. With better controls, savings could increase to or exceed 60 per cent. In this estimate, we are considering only technologies that are commercially available (at least in industrialized countries) and cost-effective. We estimate a minimum savings potential of 6 TWh/yr, realizable over the next 5 to 10 years.

Street Lighting

Sales of electricity for public lighting were 1.35 TWh in 1987/88. (Statistical Pocket Book —India, 1989). This may have increased to 1.5 TWh by 1989/90. If we add outdoor lighting equivalent to street lighting in industries, apartment complexes, etc. (which are not counted in the "public lighting" category), the total may approximate 2 TWh.

A wide range of technologies are used for street lighting in India, ranging from incandescent (the least efficient) to high-pressure sodium (one of the most efficient) bulbs. The latter produces seven times as much light per unit of electrical input (100 lumens/W vs. 15) as the former.

There are serious problems with control. Many lights stay on all day, virtually tripling energy use and reducing lamp life by two-thirds. Relatively efficient fluorescent lamps that have broken or stopped working are sometimes supplanted by incandescent lamps.

While technological alternatives can reduce the energy use considerably, institutional problems appear to dominate the efficiency of street lighting in India. In the last few years, high-pressure sodium lamps have made major inroads into street lighting. This trend should continue to replace fluorescent lamps (60-70 lumens/Watt), mercury vapor (MV) lamps (45 lumens/Watt), as well as incandescent lamps, with high-pressure sodium (HPS). Since public lighting is needed for about 10 hours per night, cost-effectiveness of high-efficiency systems is usually quite good. Geller (1990a) estimated $0.028/kWh was saved from replacing a 200-watt incandescent with a 50-watt HPS. In Brazil, there has been a trend to replace some incandescent lamps, with MV rather than HPS, presumably because the MV have much lower first cost. Since MV lamps are relatively inefficient, it would seem more appropriate to allocate the necessary capital and install HPS since the end-use efficiency immediately reduces electricity use. Photocell sensors should be used to control groups of lamps, turning them on and off according to daylight levels.

In street lighting improvement programs, lamp and ballast lifetimes must be maintained, and appropriate product improvement measures must be taken to improve life. The rated life of MV and HPS lamps in industrialized countries is 24,000 hours. Lamp life may be lower in some developing countries because of voltage fluctuations. The effect of the fluctuations may be minimized through changes in the design of the ballasts.

Savings potential. Without additional information about the distribution of types of lamps in use today, it is difficult to estimate accurately the total savings potential. However, given the relatively low penetration levels of the high-pressure sodium lamps, and the relative inefficiency of other lamp types, it is safe to assume a savings potential of at least 50 per cent, i.e., a total of at least 1 TWh/yr.

Rural Lighting

For electrified rural households, alternatives for increasing lighting efficiency are the same as for urban users compact fluorescent bulbs and fluorescent tubes provide significant savings compared to the incandescent bulb.

In lower-income developing countries, a substantial part of the rural population and part of the urban slum population does not have electric connection, and must rely on nonelectric sources of illumination. In India, these are commonly kerosene wick lamps, and some kerosene mantle lamps and candles. These provide very low levels of illumination, and are notoriously inefficient.

Table 2.5 presents a comparison of some nonelectric and electric lamps, based on laboratory measurements. Although lighting level increases from candle to kerosene wick to kerosene mantle lamp, the fuel consumption rate also increases, i.e., an improvement in light output requires an increase in fuel use rate. A shift from kerosene to electric lighting, however, results both in increased light output and reduced energy use rate. The light output increases from 40 to 730 lumens if a kerosene wick lamp is replaced by a 60-watt incandescent, but the kerosene use rate drops to half (0.032 to 0.016), even taking into account the efficiency of conversion from fuel to electricity (see notes to Table 2.5). An eightfold reduction in fuel-use rate is seen if a kerosene lamp is replaced by a 16-watt compact fluorescent. As kerosene and diesel are major fuel imports to India (and both are subsidized), significant fuel savings would result by electrifying all households, even if the electricity was generated using diesel, normally considered an expensive option. A comparison of centralized and decentralized generation alternatives is presented in Chapter 6.

Table 2.5
Comparison of Non-Electric and Electric Lamps

Light Source	Flux (lumens)	Color[a]	Luminous Efficacy (lum/W)	Fuel Consumption (kg/klum-h)	Fuel Consumption Rate (kg/h)
Candle	12	excellent	0.2[b]	0.5	0.006
Kerosene Wick	40	excellent	0.1	0.8	0.032
Kerosene Mantile	400	poor	0.8	0.1	0.040
60 W Incandescent	730	excellent	12	0.02[c]	0.016
16 W Compact Fluorescent	900	good	56	0.005	0.004

a The color quality of light sources is expressed by their Color Rendering Index; this is difficult to measure for non-electric sources, according to van der Plas.

b Luminous efficacy, lumens/W for non-electric lamps assumes a fuel to electricity conversion efficiency of 0.

c Specific fuel consumption is expressed as kg of fuel use per 1,000 lumen-hour of light (kg/klum-h). For electric sources, we again assume a conversion efficiency of 0.30 from kerosene to electricity, and express the result as an equivalent amount of kerosene.

Source: van der Plas, 1988. "Domestic lighting" originally includes additional non-electric sources—carbide lamp, butane lamp, and kerosene mantle lamp. Compact fluorescent data calculated here from data presented earlier in the chapter.

3

Residential Appliances

Residential sales of electricity in India amounted to 21.5 TWh in 1987/88 (Statistical Pocket Book—India, 1989). Extrapolating the trend, we estimate a 1989/90 consumption of 25.5 TWh. Residential demand has been rising rapidly and currently accounts for 13 per cent of total electricity sales.

A detailed analysis of electricity end uses in the residential sector is necessary in order to estimate the potential for electricity savings. Such an analysis is best accomplished by directly measuring end uses in a representative sample of households. A less rigorous approach requiring less effort is known as conditional demand analysis in which a regression technique relates total electricity use to saturation of different end-use devices (obtained from a survey) and establishes unit consumption figures.

In many developing countries, detailed end use analysis has not been carried out for the residential sector. In India, such surveys have only been conducted in a few urban areas (TERI, 1990a).

In the absence of good end-use data we calculate electricity consumed in different end uses by estimating the number of appliances in each end-use category, and taking a reference consumption figure for each appliance. We estimate electricity consumption for refrigerators, incandescent and fluorescent lamps, fans, air conditioners, and televisions. The production of electric appliances in India in recent years is given in Table 3.1.

The total number of refrigerators, air conditioners, fans, and television sets in 1990 would depend on the cumulative sales less units removed from use. If we assume a fixed appliance life of n years, then all appliances made in the last n years are still operating. We assume lifetimes of 25 for refrigerators, 15 for air conditioners, 20 for fans, and 20 for televisions. We do not have data on the production of these appliances going back that far (except for air conditioners, and even for this we have data gaps). Interpolating and extrapolating existing production data, we estimate 1990 populations as in Table 3.2.

Not all these appliances are in the residential sector. Most of the refrigerators and televisions probably are, but a significant number of air conditioners and fans are probably in commercial and

Table 3.1
Trends in the Production of Household Appliances in India

Product Units	Year								
	1975	1980	1981	1982	1983	1984	1985	1986	1987
Refrigerators									
1,000s[a]	108	278	316	350	465	646	659	591	nd
b	nd	nd	nd	330	nd	nd	668	590	680
Air conditioners									
1,000s[a]	9	26	32	30	27	29	30	40	nd
b	nd	nd	32	nd	nd	nd	32	42	37
Electric fans									
millions[a]	1.3	4.1	nd	nd	4.6	nd	5.2	4.7	nd
b	nd	nd	4.0	nd	nd	nd	6.4	6.9	6.1
c	2.1	4.2	4.7	4.8	5.2	3.9	4.8	nd	5.2
Televisions									
1,000s[a]	39	87	nd	nd	156	nd	2,480	3,000	nd

a TERI, 1990.
b Statistical Pocket Book – India 1989.
c Economic Survey 1989-90, 1990.
nd = no data

Table 3.2
Number of Appliances in India

Appliance Category	Total Number	Estimated number in Residential Sector
Refrigerators	6,500,000	5,850,000
Air conditioners	406,000	203,000
Fans	68,000,000	44,000,000
Televisions	12,000,000	11,000,000

public buildings.

If approximately 90 per cent of the refrigerators, 50 per cent of the air conditioners, 65 per cent of fans, and 90 per cent of the televisions are in the residential sector, the residential population of these appliances would be as given in Table 3.2.

To estimate their energy use we need to make additional assumptions. TERI (1990a) assumed average power input to refrigerators as follows:

165 and 235 liters	0.1 kW
286 and 310 liters	0.15 kW.

Small refrigerators are more common than large ones, and if we assume a capacity of 235 liters, the corresponding power of 0.1 kW is equivalent to an annual consumption of:

0.1 x 24 x 365 = 876 kWh/yr.

At this rate, the electricity used by domestic refrigerators is:

5.85 million x 876 kWh/yr or 5.1 TWh/yr.

TERI uses the following reference values for air conditioners:

1.0 ton	1.575kW
1.25 ton	2.0 kW
1.5 ton	2.4 kW
1.7 ton	2.65 kW
2.0 ton	3.25 kW
2.5 ton	3.5 kW
3.0 ton	4.0 kW

Domestic air conditioners are likely to be small. Thus, we assume that the average unit will draw 2.0 kW. A household survey conducted in Delhi by TERI indicated average air conditioner use of 1,687 hours a year (TERI, 1990a). If we assume 1,700 hours per year for all residential air conditioners, we may estimate total electricity use by residential air conditioners of:

203,000 x 2.0 kW x 1,700 hr/yr = 0.69 TWh/yr.

TERI assumes that ceiling fans with blades longer than 36" draw 75 watts, while those less than 36" draw 60 watts. We assume an average power input of 70 watts. TERI's Delhi survey finds 2,536 hours of ceiling fan use per year. If we assume 2,500 hours as a national average, we estimate total electricity use by residential fans of:

44 million x 0.070 kW x 2,500 hr/yr = 7.7 TWh/yr.

TERI assumes the following levels of energy use in television sets:

Black-and-white (all sizes)	80 W,
Color 14" or less	60 W, and
more than 14"	80 W.

We assume that all televisions use 80 watts. TERI's Delhi household survey revealed 3.2 hours per day of viewing for color televisions, and 3.0 for black-and-white sets. If we assume an average of 3.1 hours per day overall, our estimate for total electricity used by televisions in India is:

11 million x 0.080 kW x 3.1 x 365 = 1.0 TWh.

In Chapter 2 we showed that incandescent lamps in India use a total 16.2 TWh of electricity annually, while fluorescent lamps use 10.1 TWh. If we assume that half of the incandescent lamp electricity consumption and a fifth of fluorescent lamp electricity consumption are in the residential sector, we would estimate these components to be 8.1 and 2.0 TWh, respectively.

The breakdown of residential sector electricity use in 1990 would thus be:

End use	**TWh**	**Per cent**
Refrigerators	5.1	20.6
Air conditioners	0.8	3.2
Fans	7.7	31.2
Televisions	1.0	4.0
Incandescent lamps	8.1	32.8
Fluorescent lamps	2.0	8.1
Estimated subtotal	24.7	
Total recorded residential sales	25.5	

In estimating end uses, we arrive at a consumption subtotal that exceeds total electricity sales to the residential sector. This is not surprising, since so many assumptions were made in arriving at the individual estimates. Moreover, because of power theft, sales does not represent all of electricity use. In order to highlight the uncertainties arising from a lack of adequate end-use data, we have chosen not to adjust our assumptions to a subtotal that is below the total sales.

Our crude analysis indicates that the largest end uses of electricity in India's residential sector are lights, fans, and refrigerators. We have already considered energy-saving opportunities in lighting in Chapter 2. We will now examine the energy savings potential in fans and residential refrigerators. We will then consider end uses (air conditioners, electric water heaters, and clothes dryers) which may not appear to make up a significant share of residential electricity use at present, but could be very important in the future. Energy efficiency in these end uses must be planned for to avert problems.

Fans

In the residential sector, fans appear to be the largest electricity end-use appliance. There are no studies documenting energy efficiency improvements in fans. Gadgil (1990) makes the following estimates:

For a 30 per cent higher motor cost, one would be able to improve motor efficiency by 10 per cent (say, from 75 per cent to 83 per cent). If the motor represents 35 per cent of the total fan cost of Rs 500 ($27.78), an incremental cost of Rs 52 ($2.89) would reduce power input from 70 watts to 63 watts. If the fan is used 2,500 hours per year, annual energy savings would be 17.5 kWh. For an assumed 20-year life and a 12 per cent discount rate, the annualized cost would be Rs 6.96 ($0.39) and the cost of saved energy Rs 0.40/kWh or $0.02/kWh.

This is a very low cost of saved energy and indicates that efficient fan motors would be very cost-effective, provided we are only considering the incremental cost of the more efficient motor, i.e., the decision is made at the time the fan is purchased. In replacing an existing working fan the entire cost of the fan will need to be repaid from energy savings. An energy-efficient fan might cost Rs 550 ($30.56), at an annualized cost of Rs 76.50 ($4.25, again at a 12 per cent disount rate. If the savings are still 17.5 kWh/yr, the cost of saved energy is Rs 4.37/kWh or $0.24/kWh. This is much higher than the cost of generating electricity from most alternatives, and thus it is not cost- effective to replace a working fan with an energy-efficient one, unless the existing fan is close to the end of its useful life.

In reality, fans are not used at full speed. A manual regulator controls fan speed. It is often held that with these regulators fan energy input is not reduced when the fan is operated at low speed, and "energy-saving" electronic regulators are now widely available in India. One laboratory study compared the characteristics of conventional and electronic regulators for fan speed control (Singh and Prasanna Kumar, 1989). This study (see Table 3.3) showed that even with a conventional fan regulator (CFR) substantial energy savings are realized when a fan is operated at low speed. They additionally note that electronic fan regulators (EFR) reduce power input by 10 to 20 per cent if the fan is operated at low speed, but there is no savings at or near full speed. With the CFR the system power factor is maintained at a high value of around 0.96 at all

Table 3.3
Comparison of Conventional and Electronic Fan Speed Regulators

Step Position of Fan Regulator	Conventional Regulator				Electronic Regulator			
	Volts (V)	Current (mA)	Power (W)	Power factor	Volts (V)	Current (mA)	Power (W)	Power factor
1	138	157	37	0.98	138	192	29	0.63
2	168	189	44	0.97	168	221	37	0.69
3	188	208	48	0.96	187	239	44	0.77
4	212	234	54	0.96	214	254	54	0.89
5	240	251	58	0.96	237	253	58	0.96

Notes: The input voltage was kept constant at 240 V.
The fan speed was measured and the electronic regulator (which is continuously variable) was set to equal the speed at the 5 settings of the conventional regulator.
Source: Singh and Prasanna Kumar, 1989.

speeds; when the EFR is used, however, the system power factor drops in reduced speed operation. Moreover, the line current draw with the EFR is substantially higher at low fan speeds compared to the CFR. This would increase transmission and distribution losses. The EFR introduces harmonics into the line current, which can reduce fan life, hinder the proper operation of other electronic equipment, and damage the distribution transformer. If all electronic regulators work on the same principle, as is suggested by Singh and Prasanna Kumar, it is doubtful whether they should be recommended as an energy-saving measure. Given the importance of fans as an end user of energy, alternative, inexpensive speed control devices need to be developed that permit speed control without distorting the wave form and maintaining high power factor.

The reduced power input at reduced speeds, even with the conventional regulator, suggests that fan power input is less than the 70 watts assumed in the end-use breakdown above. If the average fan input were 54 watts, instead, corresponding to the "4" setting with the conventional regulator, the total energy consumption in fans would be 5.9 TWh/year instead of our earlier estimate of 7.7.

Thus, the main energy-saving option available for fans is through energy-efficient motors. If all new fans had energy-efficient motors, a 10 per cent energy savings would be realized on all new fans sold. If we assume the average fan power input is 54 watts as discussed above, and 2,500 hours of annual operation, electricity savings corresponding to sales of 5 million units a year would be

5 million x 0.054 kW x 2,500 hr/yr = 0.68 TWh for the first year.

In the next year these fans will save another 0.68 TWh; there will also be savings from the following year's sales. If the sales are steady at 5 million a year, and fans last an average of 20 years, the annual savings would steadily increase, to 20 x 0.68 or 13.5 TWh in the 20th year relative to the case where there was no efficiency improvement.

Conservation potential for India. A year 2000 conservation potential may be calculated as follows, assuming an estimated 44 million fans in India's residential sector drawing 70 watts and operating 2,500 hours a year consumed 7.7 TWh in 1989-90. If 65 per cent of the 5 million fans sold each year end up in the residential sector (i.e., 3.25 million), and 36.25 million of the present fans in the residential sector are functioning in the year 2000, consumption in that year would be:

(11x3.25+36.25) million x 70W x 2500 h = 12.6 TWh

if there were no efficiency improvement.

As noted above, replacing a working fan with an energy efficient one is not cost effective (cost of saved electricity $0.24/kWh). Although electronic regulators currently on the market could reduce fan power, these produce undesirable power line harmonics and probably should not be promoted on a large scale. We assume that from 1994 on all new fans sold have energy efficient motors, improved aerodynamics, and acceptable speed controllers so that average power input is reduced to 54 W. From 1990 to 1993, we assume that there is no improvement in efficiency. For the same number of total fan sales and the same proportion in the residential sector, we have consumption in the year 2000 as:

[(36.25 + 4 x 3.25) million x 70 W + 7 x 3.25 million x 54 W] x 2500 h = 11.7 TWh.

Thus, a year 2000 conservation potential is 12.6 - 11.7 or 0.9 TWh.

Refrigerators

After lights and fans, refrigerators are India's next largest electricity user. We calculate that household refrigerators consume 5.1 TWh of electricity, using TERI's estimate that an average Indian refrigerator uses 876 kWh energy per year. This nominal consumption may be compared with reference values of other countries, based on measured performance, as shown in Table 3.4. We find that the energy use in a two-door Japanese model of comparable size in 1973 was somewhat higher 955 kWh/yr. The average European refrigerator at the time used much less energy but this cannot be readily compared since the European units did not include a freezer, and because test procedures in different countries are different. Nevertheless, it is clear from the Japanese and European data that average energy use dropped drastically since the Arab oil embargo in 1973. The best European model sold in 1988 used only one-sixth as much electricity as the average in use in 1973. Advanced prototypes could reduce this energy use by another 40 per cent.

The Brazilian case is perhaps more relevant to what can be achieved in a developing country. In 1984, Brazil initiated a refrigerator efficiency program consisting of efficiency testing and labeling. In 1986, the average one-door model sold (size range 250-300 liters) used only 540 kWh per year. In 1989, one manufacturer introduced a model that used only 336 kWh per year. The electricity consumption of residential refrigerators has been reduced through improved insulation, more efficient motor-compressors, modified design of the refrigeration cycle, and reduced the capacity of resistance heaters. Because of improvements made during 1987/89, Brazilian refrigerators saved approximately 0.38 TWh of electricity in 1989, equivalent to the power supplied by 100 MW of hydroelectric capacity (Geller, 1990).

Encouraged by the progress achieved, and following the additional savings potential realized in some industrialized countries, in 1990 Brazil proposed minimum efficiency levels for refrigerators through agreements between the government and manufacturers, requiring that by 1993 refrigerators in each category consume no more than the most efficient model available in 1989 (Marques, 1990). The proposed agreement calls for efficiency improvements of 5 per cent a year from 1994 to 1998. These additional savings could lead to 1998 models using only half as much as electricity as those sold in 1990 (Geller, 1990).

Table 3.4
An International Comparison of Electricity Use in Residential Refrigeration

Country	Capacity (liters)	Special Features	Year	Consumption[a] (kWh/yr)	
Japan[b]	170	2-door	1973		955
			1982		372
			1984		324
Europe[c]	200	no freezer	1973	use	550
			1988	use	350
			1988	sold	270
			1988	best	90
				advanced	50
Brazil[d]	250-300	1-door	1986	sold	540
			1988	sold	454
			1989	best	336

a The values from different countries are not strictly comparable since test procedures followed in different countries are not the same. "Use" refers to the stock of units in use in a given year; "sold" refers to the energy use of the average unit sold in a given year; "best refers to the energy use of the most energy-efficient model sold in a given year.
b Nikura, 1986.
c Norgard, 1989.
d Marques, 1990.

The US experience showed that improvements in refrigerator energy efficiency were very cost-effective (low cost of saved energy) but that implicit discount rates for investment in energy-efficient refrigerators were high-consumers overwhelmingly favored the lower purchase cost models. To correct this market imperfection in refrigerators and other residential appliances, in 1986 the US Government promulgated nationwide minimum appliance standards. Earlier, the state of California had set minimum efficiency standards for certain appliances. Starting in 1993, the US called for tightening of efficiency standards of refrigerators and other appliances. The analysis of additional costs associated with various levels of efficiency improvement can be used to exhibit their cost-effectiveness. Refrigerators used in the US are much larger than those commonly sold in developing countries. In Table 3.5 we show the results of the analysis for a 15.1-cubic foot (428-liter) manual defrost refrigerator, as the model closest in characteristics to those sold in developing countries. Since this unit has about twice the volume of the typical Indian refrigerator, we add a column normalizing consumption to liters of capacity for comparison.

Table 3.5
Energy Use and Additional Manufacturing Cost of a 15.1 Cubic Foot (428 liter) Manual Defrost Refrigerator in the USA[a]

Level	Description	Average Energy Use (kWh/yr)	Normalized Use (kWh/yr-lit)	Manufact. Cost (US$/unit)	Marginal CSE[a] ($/kWh)
0	Baseline	619	1.45	154	na
3	Plus foam refrigerator door	531	1.24	155	0.001
5	Plus 2" door insulation	505	1.18	159	0.020
7	Plus 2.5" back and side insulation	464	1.08	169	0.031
8	Level 5 + 3" back and side insulation	1.02	176	0.026	
9	Level 5 + evacuated panels	325	0.76	201	0.029

a The manufacturer's cost of the different models assumes that ozone-depleting chlorofluorocarbons (CFCs) are not available as foam insulation blowing agents. Instead, they consider the use of HCFCs that have much lower ozone depletion potential but do not provide the same level of insulation per unit thickness. If no CFC restrictions were present, an energy consumption level of 253 kWh/yr could be realized at a manufacturer's cost of $209 (DOE, 1989, Table 3.25).

Marginal CSE is the cost of saved energy based on the extra cost of a given improvement relative to the characteristics specified on the previous line. The extra cost is annualized using a discount rate of 12 per cent, assuming a useful life of 25 years.

Source: DOE, 1989.

Clearly the manufacturing capability and additional cost are not the same in all countries. Note however that the analysis in Table 3.5 only considers insulation improvements, and that all the technologies (except for the use of evacuated panels) already exist or can be readily introduced. Significant energy savings are possible at a cost of saved energy at or below $0.03/kWh. The most efficient model in Table 3.5 would use only 325 kWh/yr, less than 40 per cent of the energy consumption assumed for Indian refrigerators (876 kWh/yr), even though US refrigerators are about twice as large. Thus efficiency improvements by a factor of about five should be realizable in India over the next five to seven years, provided appropriate steps are taken in this direction. A realistic target might be a specific energy consumption of 1 kWh/liter for refrigerators manufactured in, say, 1998, equal to the standard recently proposed in Brazil for that year (Geller, 1990).

Experience in various countries has shown that the mere

existence of energy-efficient models that are cost-effective to own and operate does not mean that consumers will make economically rational decisions. Even mandatory labeling of the energy efficiency on a bright, easy-to-read label (in place since May 1980 in the US), did little to improve the purchase of energy-efficient refrigerators. For energy-efficient refrigerators to penetrate the market, the efficiency of new refrigerator models must be tested and either minimum efficiency standards must be set (as in the US) or manufacturers must agree on efficiency protocols (as in Brazil).

In India, the energy consumption of refrigerators is tested as part of specifications covering refrigerators that carry the Indian Standards Institution label (IS, 1985). The label is considered to be a sign of product quality, and most (if not all) refrigerators manufactured in India carry this label. Thus the energy consumption of refrigerator models sold in India is known, although no compilation of these values is currently available. The first step of an energy efficiency program would be to make such a compilation and gauge the current status. This needs to be supplemented by independent measurements of energy consumption both in the laboratory and in actual use.

If the average new refrigerator sold uses only 200 kWh per year instead of our assumed 876 kWh per year, the annual savings per refrigerator would be 676 kWh per year. If annual sales of refrigerators are 600,000, total energy savings would be: 600,000 x 676 kWh/yr = 0.41 in the first year, twice that in the second year and so on. Since refrigerators are expected to last 25 years, a rate of refrigerator sales of 600,000 per year would mean that electricity savings would steadily increase to 25 x 0.41 or 10 TWh/yr.

Conservation potential for India. A year 2000 electricity conservation potential may be calculated assuming that,

1. in 1989-1990, 5.85 million refrigerators were in the residential sector, each consuming 876 kWh per year, or a total of 5.1 TWh a year;
2. 5.4 million of these survive to the year 2000; and
3. 90% of 0.6 million new refrigerators sold each year end up in the residential sector, and all these consume 876 kWh/yr.

Without energy conservation, total consumption in the year 2000 would be:

$$(5.4 + 11 \times 0.9 \times 0.6) \text{ million} \times 876 \text{ kWh} = 9.9 \text{ TWh /yr.}$$

If there is no change in refrigerator efficiency through the year 1994, and the average consumption of new refrigerators manufactured from 1995 on is 200 kWh/yr, the total year 2000 consumption would be:

> (5.4 + 6 x 0.9 x 0.6) million x 876 kWh + (5 x 0.9 x 0.6) million x 200 kWh = 8.1 TWh per year.

Thus the year 2000 conservation potential in residential refrigerators is 1.8 TWh, given the above assumptions.

Other Appliances

Although lights, fans, and refrigerators account for most electricity use in Indian households, some other appliances are worthy of mention.

Water heaters. A chemical or nuclear fuel is typically burnt to raise a fluid to a very high temperature and thereby generate electricity in a capital-intensive power plant. Electricity suffers transmission and distribution losses on its way to residences. To use this high quality energy source to heat water from 25 to 60°C is thermodynamically a poor use of energy.

In many industrialized countries and some developing countries, a fuel (such as natural gas or fuel oil) is burned in homes to heat water. This is thermodynamically not much better but at least it avoids the capital cost of the power plant. Where a fuel is available for heating water, it is cheaper than heating water with electricity. Unfortunately, in petroleum-poor developing countries, a water heating fuel is not available, and households often heat water with electricity. Although domestic water heaters are rare in India, most run on electricity. Several alternatives to the standard electric water heater are worth considering: solar water heaters, heat pump water heaters, and piped natural gas or producer gas water heaters.

Solar water heaters are the most inexpensive of the many ways to use solar energy. Relatively simple materials are needed to raise water temperature to 60°C. Since winter, when hot water is most needed, is the dry season in most of India and there is little cloud cover, solar water heating is ideally suited to the climate. The Indian Government has long recognized the merits of solar water heating and provides a large subsidy towards the installation of solar water heaters. Nevertheless, as of 31 January 1989, only 3,025 domestic solar water heaters had been installed (TERI, 1989b). Many designs exist, and there are more than a handful of manufacturers in India. In some countries, such as Israel, solar water heaters are widely disseminated. Why they are not widely disseminated in India is not well-documented, but needs to be resolved.

Heat pump water heaters use the heat pump principle (as in a refrigerator or air conditioner, but in reverse) to heat water, requiring only about one unit of electricity to heat three units of water. Heat pumps are over three times more efficient than ordinary electric resistance water heaters. In the US, they have been shown to be as effective as solar water heaters (heat output per unit electricity expended), but at considerably lower cost. Nevertheless they are expensive, costing about $1,500 (in the US) and it is not clear that they can compete economically with ordinary electric water heaters or with solar water heaters (which for various climatic reasons are considerably cheaper in India). The hot water demand in a typical Indian household is small compared to that in an industrialized country, and it remains to be seen if an inexpensive heat pump water heater can be designed for the low demand.

In some parts of India, natural gas is being developed as fuel. In the Arabian sea, enormous quantities of gas associated with oil wells were flared until recently, when some large pipeline projects brought the gas to possible load centers. Natural gas is an ideal cooking fuel but piping it to houses just to satisfy cooking demand is likely to be uneconomical. If the gas were to be used for cooking and water heating, then the economics may change, making it accessible for both end uses. In some cities, such as Calcutta, a coal-gas-based pipeline connected to households has been in place for several decades. In recent years, the supply of gas to Calcutta has ceased pending major repairs in the gas plants. Once restored, it may be both a convenient cooking and water-heating fuel.

Washing machines. The demand for electrically heated water for washing machines is increasing indirectly and rapidly in India. Washing machines, used in the industrialized countries for many decades, were not popular in India because the appliance-purchasing classes had access to cheap labor for handwashing clothes. Recently, however, a number of manufacturers have entered the washing machine market, mostly with foreign collaboration. Sales in 1990 are expected to be "nearly 175,000 automatic and semi-automatic washing machines" (Rahman, 1990). One industry representative projects sales to reach 500,000 annually. We do not know what fraction will include electrically heated water (EHW). If we consider a steady sales rate of 200,000 EHW units a year, and consider that the average unit uses 400 kWh per year (the same as the average washing machine in use in Denmark in 1988 (Norgard, 1989) and assume a life of 15 years, the total electricity demand for clothes washing would be:

$$200{,}000/\text{yr} \times 400\ \text{kWh/yr} \times 15\ \text{yr} = 1.2\ \text{TWh/yr}$$

in the steady state assumed, with a stock of 3 million EHW washing machines.

Norgard (1989) notes that the best washing machine available in Denmark in 1988 used only 240 kWh per year while an efficiency-advanced prototype requires only 115 kWh per year. These consumption figures correspond to a 4-kg washing machine used for 200 washings a year, typical of Denmark.

Most electricity used in the conventional washing machines goes to heating water. One electricity-saving measure would be to use water heated in a household water heater based on a different fuel, as has been suggested earlier (and is normal practice in the US and Canada).

Far larger savings are possible using cold water detergents. These are commonly available in industrialized countries and many households in the US wash with cold water (8 to 15°C). Moreover, enzymatic detergents dissolve proteins and can eliminate much of the mechanical action in the washing process. Lukewarm water speeds up the enzyme process while hot water ruins it. The warmer ambient water temperatures typical of tropical developing countries are best suited to this type of detergent. Other detergents capable of dissolving fat in cold water also exist. Clearly, substantial electricity savings in clothes washing will depend mostly on detergents that permit cold water washing.

The recent trend in washing machine purchase in India is cause for concern. Top-loading machines (typical of North American designs) require more water, and hence more hot water and electricity than front-loading models (typical of Europe). Top-loading machines have captured the lion's share of the Indian market (Rahman, 1990). This is the result of the lower purchase price of the top-loading machines and is yet another example of how consumers are essentially driven primarily by purchase cost considerations, rather than by life-cycle cost. (The complete absence of information about operating cost does not help either.) A forward-looking industrial and energy policy should promote advanced detergents, set minimum efficiency standards, or prohibit the sale of all machines that heat water with electricity. Otherwise, we will be throwing a terrawatt hour out with the wash water.

Air conditioners. While not currently a major end user of electricity in India, the use of air conditioning expands with increasing standards of living, and is significant and growing in moderately upper income developing countries and among the upper income strata of all developing countries. An increase in overall income, particularly in the upper strata, can lead to a major increase in the demand for air conditioners. Since air conditioners are energy-intensive, their energy efficiency has a significant impact on future demand. It should be noted that consumers are particularly reluctant to invest in efficiency of air conditioners. In the US, purchases of room air conditioners were driven entirely by puchase cost, even after energy consumption labels had been in place for years. Ironically, energy efficiency improvements are very cost-effective. Gadgil estimates that at an additional cost of Rs 2,000 ($111) the seasonal energy efficiency ratio (SEER, the energy index used to rate air conditioners in the US) can be increased from a nominal value of 6 to 9. If there are no additional taxes for the extra cost for energy efficiency, the cost of saved energy is excellent—$0.04/kWh (Rs 0.72)—(see Table 3.6), but with a 25 per cent import charge and a 100 per cent excise tax over that, the cost of saved energy increases by a factor of 2.5 to $0.10/kWh (Rs 1.85/kWh), which is currently more than the (subsidized) price of electricity to the residential sector, and comparable to the long-run marginal cost of supplying electricity (see Chapter 6).

Table 3.6
Cost Effectiveness of a More Efficient 1-Ton Room Air Conditioner

	Extra Cost Rs (US$)	Annualized Capital Cost Rs ($)	Electricity Savings (kWh/yr)	Cost of Saved Energy Rs/kWh ($/kWh)
Rs.	2,000	293.65	397	.74
$	(111)	(16.31)	397	(.04)

Notes: Air conditioner SEER increases from 6 to 9.
The additional cost does not include any taxes.
Discount rate of 12 per cent, and a life of 15 years are assumed.
Source: Gadgil, 1990.

This example illustrates the nature of market imperfections which is by no means unique to India: consumer indifference to energy efficiency, electricity price subsidy, and a tax structure that reduces the cost-effectiveness of energy efficiency improvements. The result is that electricity will be wasted by wealthy households that own air conditioners and pay a low rate for electricity. The electric utilities will then try to compensate by building more expensive power plants, increasing budget deficits, external debt, and the load on the environment.

Against such a gross market imperfection, minimum efficiency standards are essential to keep inefficient equipment off the market.

Energy savings potential for India. For the year 2000, potential electricity savings for air conditioners may be estimated assuming that the average residential air conditioner consumes 2.0 kW for 1,700 hours of operation per year. All air conditioners (203,000 units) consumed 0.69 TWh in 1989-90. Air conditioners have a shorter life than refrigerators and perhaps 125,000 of the currently operating units will survive to the year 2000. Additionally we asume that 20,000 new units will enter residential service annually till that year. Without efficiency improvement, consumption in the year 2000 would be:

$$(125{,}000 + 11 \times 20{,}000) \times 2000 \text{ W} \times 1700 \text{ h} = 1.2 \text{ TWh.}$$

Assuming that energy-efficient air conditioners consume a third less energy (e.g. with a seasonal energy efficiency ratio of 9 instead of 6) and that they can be introduced to meet all demand in 1994 (with no efficiency improvement until then), the year 2000 energy consumption (with conservation) would be:

$$[(125{,}000 + 4 \times 20{,}000) \times 2000\text{W} + (7 \times 20{,}0000 \times (0.667 \times 2000\text{W})] \times 1700 \text{ h} = 1.0 \text{ TWh per year.}$$

The year 2000 energy conservation potential in residential air conditioners is thus 1.2-1.0 or 0.2 TWh a year.

Conclusion

We have reviewed three "residential" appliances—fans, refrigerators, and air conditioners—which are also found in other sectors. Above, we have assumed that 65 per cent of fans, 90 per cent of refrigerators, and 50 per cent of air conditioners are located in the residential sector, and have estimated year 2000 energy savings potential. We may also estimate the conservation potential in these appliance categories considering all the appliances irrespective of the sector in which they are used. The results are summarized in Table 3.7.

Table 3.7
Year 2000 Electricity Consumption with Frozen Efficiency and Conservation Potential

	Consumption 1989-90	Consumption 2000[a]	Conservation Potential, 2000
Refrigerators	5.7	11.0[b]	2.0[c]
Fans	11.9	19.3[d]	1.4[e]
Air conditioners	2.1	3.5[f]	0.5[g]

a frozen efficiency

b Of the 1990 population of 6.5 million refrigerators, 6 million are expected to survive to the year 2000 (since most of the present stock was added in the 1980s). To this we add the current production of 600,000 a year for 11 years. We assume that all refrigerators consume 876 kWh/year.

c We assume that from 1995 on the average refrigerator consumes 200 kWh/yr, so that we have (6 + 0.6 x 6) million refrigerators consuming 876 kWh/yr and 0.6 x 5 million consuming 200 kWh, for a total consumption of 9.0 TWh, so that the conservation potential is 2.0 TWh.

d Assuming that India's 68 million fans consume 70 watts on average and are used 2500 hours a year, present consumption is 11.9 TWh a year. With 5 million sold a year, and say 55 million of the present stock remaining, year 2000 consumption would be (110 million x 70 W x 2500 h) =19.3 TWh.

e We note that replacing a working fan with an energy efficient one is not cost effective (cost of save electricity $0.24/kWh). Although electronic regulators currently on the market could reduce fan power, these produce undesirable power line harmonics and probably should not be promoted on a large scale. We assume that from 1994 on all new fans sold have energy efficient motors, improved aerodynamics, and acceptable speed controllers so that average power input is reduced to 54 W. From 1990-93, business as usual. Fan sales 5 million a year. Thus year 2000 consumption (with conservation) = ((55 million + 4 x 5 million) x 70 W + 7 x 5 million x 54 W) x 2500 h = 17.9 TWh, i.e. a conservation potential of 19.3 - 17.9 or 1.4 TWh

f There were 406,000 air conditioners in India in 1990. If we assume the average unit draws 3 kW and are used 1700 h a year, total consumption is 2.1 TWh. Sales have increased considerably and we may expect sales to average 40,000 units a year until the year 2000. Air conditioners have a shorter life comapared to refrigerators and perhaps 250,000 of the present stock will survive to 2000. Thus F.E. consumption in 2000 would be (250,000 + 11 x 40,000) x 3000 W x 1700 h = 3.5 TWh.

g. Assuming that energy efficient air conditioners use a third less (e.g. by increasing seasonal energy efficiency ratio from 6 to 9) can be introduced in 1994, with business as usual until then, the year 2000 energy consumption (with conservation) would be ((250,000 + 4 x 40,000) x 3000 W + 7 x 40,000 x 2000 W) x 1700 h = 3.0 TWh, i.e. a conservation potential of 0.5 TWh.

Table 3.8
Conservation Potential of Refrigerators, Fans, and Air Conditioners in Commercial Buildings, Year 2000

Appliance	Conservation potential, 2000 (Terawatt-hours)
Refrigerators	0.2
Fans	0.5
Air conditioners	0.3
Total	**1.0**

If we subtract the consumption and savings potential in the residential sector and assume that the remaining use is all in commercial buildings, the conservation potential of these appliances in commercial buildings is as shown in Table 3.8.

Thus, we estimate an energy conservation potential in commercial buildings of 1.0 TWh for end uses other than lighting.

4

WATER PUMPING

Energy is used for pumping water in lift irrigation, pumping water and sewage in urban areas, and pumping for a variety of industrial tasks. In India, most of the energy used in pumping is for irrigation. Agricultural pump sets may be fueled by electricity or diesel, but electric pump sets comprise virtually all of the sales in the agricultural tariff category.

Electricity sales to agriculture was 23.42 TWh in 1985/86, and was estimated to be 28.22 TWh in 1986/87 and 34.81 TWh in 1987/88 (Statistical Pocket Book — India, 1989). Following this trend, 1989/90 sales would be at least 43 TWh. As we see below, this estimate is too high.

Electricity sales to agriculture showed the highest rate of increase over the last two decades — from 10.2 per cent of the total in 1970/71 to 23.8 per cent in 1987/88. Residential use experienced the second highest increase, from 8.8 per cent to 14.7 per cent of the total, over the same period. Commercial electricity use remained roughly steady, while the shares of industry and railways have declined.

In 1988/89 there were 7.6 million electric pump sets (TERI, 1989b). (The number of diesel pump sets has not been estimated after 1983/84, when there were 3.1 million. If we assume there are the same proportion of diesel and electric in 1988/89 as in 1984/85, the total population of pumps would be 12.1 million, the same as estimated by Tolpadi [nd].)

The 1985/86 total electric consumption of 23.42 TWh is equivalent to a consumption per pump set of 3,807 kWh per year, a number that has been slowly increasing over the years. We estimate a 1989/90 average consumption of 4,000 kWh per pump set. For an estimated 8 million electric pump sets in 1989/90, the total electricity used in pumping would be 32 TWh, not the 43 TWh noted above.

A number of large-scale experiments have been carried out to empirically determine the energy savings potential in agricultural pumping. The best documented cases are in Gujarat, where four levels of retrofits of increasing cost (Table 4.1) were considered for existing electric and diesel pumps. Costs, savings, and the cost of saved energy for electric pump sets are summarized in Table 4.2.

Table 4.1
Energy-Saving Retrofits (or Rectifications) to Electric Pump Sets

Retrofit Level	Description
R1	Replacement of high friction suction line and foot-valve of centrifugal pump by low friction suction line and valve
R2	R1 plus similar change in delivery line
R3	R2 plus replacement of inefficient pump by an efficient pump
R4	Complete replacement of the existing pump, motor, piping system, etc. by a new efficient system, i.e., R2 plus a new pump set

Source: Patel, 1988, Table 1.

Table 4.2
Economics of Electricity Conservation In Irrigation Pumping

Retrofit Level	Investment ($ [1990])	Savings (kWh/yr)	Savings (Per cent)	Cost of Saved (Energy, $/kWh)
R1	34.11	1,513	22	0.004
R2	80.94	2,370	29	0.006
R4	594.61	9,582	47	0.011

Notes:

- Original data refer to projects conducted over a number of years. The costs in Rupees are assumed to be average values at year end 1984. They are updated to end-1990 as in the rest of this report using a 7 per cent inflation rate over the six-year period. Finally these figures are converted to $(1990) at Rs 18/$.
- Savings are the average values per installation.
- A 10-year life and a 12 per cent discount rate are assumed to calculate capital recovery factor and the cost of saved electricity.
- Sample sizes for the three retrofit levels evaluated are 1200, 1200, and 300 respectively. Average initial consumption levels were substantially different for the three retrofit sets – 6,892, 8,157, and 20,198 kWh/yr, respectively.

Source: Patel, 1988.

All three levels of retrofit agricultural pumps are seen to be highly cost-effective from a societal perspective, with a cost of about \$0.01/kWh saved, much less than the \$0.09/kWh or more for supplying electricity from new coal or nuclear power plants. The tariffs prevalent for irrigation pumping in India are heavily subsidized, often no more than a flat fee per kW of connected load, irrespective of the energy used. The average price in 1988/89 for agricultural customers was only Rs 0.16/kWh (\$0.01/kWh), compared to an average cost of generation and supply of Rs 0.91/kWh (\$0.05/kWh) (CMIE, 1990). Only in the R4 case, where the existing motor is replaced by a smaller one, is there a reduction in load.

In the R4 experiment reported in Table 4.1, the average load reduction was 4.9 kW. The average investment of \$594.61 works out to a unit cost of \$121/kW. Patel (1988) proposes a subsidy of Rs 500/kW "returned," and shows that this would be far more cost-effective than the subsidies given by the Indian Government for solar and wind-powered pumping.

Savings Potential

The Gujarat experience has shown that almost 50 per cent savings is achievable with the complete R4 retrofit at a cost of saved energy of about \$0.01/kWh, much less than that of electricity supply. These savings were observed in pumps that consumed more electricity than the national average which we earlier estimated to be 4,000 kWh/yr (see last note, Table 4.2). If we assume that only 35 per cent savings is possible on average, then for a 1989/90 consumption in agricultural pumping of 43 TWh, the total savings potential would be 15 TWh/yr. The electricity generation avoided would be substantially larger if we consider the large transmission and distribution losses for the highly dispersed rural electric load.

Implementation Strategy

Tolpadi proposed detailed implementation strategies for new and existing water pumping systems. We summarize these here.

New pumps. The Bureau of Indian Standards has a number of standards that set minimum efficiencies for the various elements of the pumping system. Although higher efficiencies are possible, a convenient starting point would be to make these standards mandatory. This can be implemented through the various lending institutions, departments providing agricultural subsidies and loans, etc. For instance:

- electricity boards may wish to subsidize part or all of the costs of the efficiency improvement, and/or

- loans and subsidies could be limited to energy-efficient systems.

Priority may be given to new electric connections for energy-efficient systems.

Demonstration and publicity must be organized at the smallest administrative levels to inform people at all levels—from manufacturers to lenders to users—involved in the construction, marketing or use of pumps.

The Indian Pump Manufacturers Association (IPMA) is in the process of setting up an Indian Hydraulic Research Center. Some of its objectives concur with the requirements of improving the capabilities for designing and manufacturing efficient pumps (IPMA, 1990). These include:

- making research and development facilities available to the pump industry and pump users,
- assisting small-scale pump manufacturers to update designs of their agricultural pumps,
- providing independent testing for pump performance,
- assisting in the formulation and review of Indian Standards on pumps, and
- developing a pump technical library and documentation center.

State electricity boards and agricultural departments need to have trained personnel to implement and verify energy conservation goals. Training is essential at all levels. An existing organization that has sponsored many successful experiments and evaluations could be responsible for the task. In India, the National Bank for Agriculture and Rural Development appears most suited to organize the implementation, with technical support from the Petroleum Conservation Research Association and Rural Electric Corporation which have been involved in diesel and electric pump sets, respectively, as well as IPMA. Another important issue for new pump sets is the choice of energy source—see below *Alternatives to Electric Pumping.*

Existing pumps. As noted above, there were some 8 million pump sets operated by electricity in 1989/90.

Tolpadi notes that replacement of the foot valves is the single most cost-effective investment, saving 8 to 12 per cent at a cost of Rs 150 (about $9). This gives a $0.004/kWh cost of saved energy, assuming a 10 per cent reduction for the average pump set energy consumption of 4,000 kWh/yr. However, while energy is saved, no power is saved, and at current tariff structures, there is unlikely to be an incentive for this retrofit. Thus, Tolpadi proposes free installation of a million replacement foot valves, an exercise that has been carried out by many State electric boards.

Our calculation of cost-effectiveness of the higher level of retrofits (RI, R2, and even R4 in Table 4.2) shows that they are all very cost-effective compared to the cost of providing electricity, and in fact more so than many other conservation opportunities. We therefore believe that sufficient priority should be given to allow the full retrofit (Table 4.2, R4) of as many pumps as is feasible given infrastructural constraints (number of trained people available, speed of training additional personnel, administrative bottlenecks, etc.). The retrofits may be carried out by cooperatives, the Rural Development Corporation, pump manufacturers, etc. Tolpadi proposed 150,000 R4 retrofits during the first five years. Given that there are at least 8 million electrically powered pumps, with plan-targeted additions of 600,000 a year between 1985/90, the rate of retrofits (rectifications) needs to be increased rapidly (over five years?) to at least 150,000 a year. Even at that rate it will take 50 years to fix all the existing electric pump sets. Priority should be given to larger energy users, where the cost-effectiveness will be higher, and to systems requiring repairs or modifications.

Alternatives to Electric Pumping

We noted above that diesel pumps are widely used for water pumping in India. At private costs based on rural electricity tariffs and diesel prices (inexpensive relative to petrol), electric pump sets are much less expensive than those operated on diesel. However, comparison based on shadow prices reflecting social costs reveals that the two alternatives cost virtually the same, with diesel actually being somewhat less expensive (Table 4.3). The relative abundance of electric pump sets has no doubt been influenced by economics as perceived by users. Bhatia (1990) pointed out that dual fuel biogas/diesel engines have lower social costs than diesel engines for agricultural pumping. Thus, if heavy subsidies for agricultural pumping with electricity cannot be removed for political reasons, nonmarket mechanisms must be implemented to discourage new electrical connections. Note that the energy conservation opportunities identified above apply equally to diesel engines and to biogas/diesel engines.

Producer gas/diesel dual fuel engines and wind pumps are other alternatives. Bhatia (1990) found the producer gas option to be more expensive than diesel, but the relative costs may be different based on recent commercialization of improved gasifiers and increased diesel prices following the 1990 Gulf crisis. The economics of

Table 4.3
Cost Comparison of Electric and Diesel Pump Sets

	Capital Cost, US$	Operating Cost $/yr	Annualized Life Cycle Cost $/yr
Using Private Costs			
Electric	373.33	76.67	142.74
Diesel	510.00	615.10	705.36
Using Social Costs			
Electric	3,060.00	214.17	755.74
Electric	543.33	623.78	719.94

Notes:

- Ramesh and Natarajan assumed a shadow price of Rs 0.75/kWh to calculate social costs. The marginal cost of supplying electricity is considerably higher – at least Rs 2/kWh. Thus the real social cost of electric pump set operation is even higher.
- They used a 10 per cent discount rate and calculated the net present value for the four alternatives. We have used a 12 per cent discount rate and computed the annualized life cycle cost instead, for consistency with other analyses in this report.
- We have converted their Rupee values to US$ at an exchange rate of Rs 15/$, so that the results above are in 1987 dollars. Diesel and electric prices have not increased at the rate of general inflation, and we have not attempted to bring the comparison up-to-date in order to keep the same basis as the authors. Our conclusions are identical to theirs.

Source: Ramesh and Natarajan, 1987, Table 5.

windmills is very site-specific, depending on the availability of wind. One analysis found the windmills to be marginally more cost-effective than diesel pumps but warned that the windmill evaluated "is at best suitable for extremely small and marginal farmers" (Srivastava and Sharan, 1987). Both alternatives deserve continuing development.

Implications for Other Countries

The results for India should be generally valid in other Asian countries. The savings potential will depend on the initial friction losses, and motor and pump efficiency. A major difference is that there are relatively few electrically operated pumps in agriculture, even in neighboring Pakistan and Bangladesh (see Table 4.4). Thus even if the savings per pump were the same, the total savings potential would be much less.

Other Pumps

Irrigation is of course not the only place that pumps are used. Water supply and sewage pumping consumes a lot of energy. In India, sales to these end uses was 2.6 TWh in 1987/88, and was projected to be 3.0 TWh in 1989/90 (Statistical Pocket Book—India, 1989). Here, the opportunities for electricity savings through better management are very large. Many public water supply points in

Table 4.4
Electrically Operated Pumps in Some Asian Countries

Country	Year	Number of Pumps	Pumps per 1000 sq km	Pumps per million Population
Bangladesh	1983	9,200	64	129
India	1983	5,130,000	1,620	7,502
Pakistan	1982	76,397	95	911
Philippines	1981	750	3	16
Thailand	1982	375	2	8

Source: Ramesh and Natarajan, 1987.

some urban areas do not even have faucets, but an open pipe through which water pours whenever the municipality is pumping water. In the dry season there may not be much wastage, as people line up to fill their buckets, but at other times water may flow unchecked for long periods of time. Even if water is not scarce in these areas, the cost of pumping surely is considerable. We have not found quantification of these losses. Losses should be quantified as the first step at energy management, even before considering whether the pipe friction may be reduced or a new energy-efficient motor is in order.

Pumps are used in households and offices to raise water from street level to the roof, from which it is distributed by gravity throughout the building. Here, too, management must precede mechanical improvement. Better management might be achieved by installing controllers that shut off the pump whenever the roof tank is nearly full, thereby avoiding losses through spillage.

Water conservation has much to recommend itself for its energy-saving potential. Toilet flushes represent a major use of water. Low-flow toilets (which work as well or better than conventional ones) not only consume less water per flush but are likely to be more reliable. In Mexico City, following a program of testing, designing and developing low-flow toilets, a large part of public and other high-use toilets were replaced. This program may be emulated in other countries. Low-flow toilets are available in India, and could be required for all new construction.

End-use studies of water flow in buildings and its energy implications are urgently needed in all developing countries.

The industrial sector is another large user of pumps. Here, apart from savings from more efficient motor drives (discussed in the next chapter) and reduced friction losses, are other opportunities for conservation. Where variable flow rates are needed, adjustable speed drives are an alternative to pump throttling. The pumping system needs to be analyzed and optimal combinations of motor, pump, piping, and controls defined. Baldwin (1989) shows how a combination of measures could reduce primary energy use by 63 per cent.

5

Industrial Motors

The industrial sector accounts for the largest share of electricity use in India. In 1987/88, industries purchased 77 TWh from electric utilities. Although up-to-date information on self-generation of electricity by industry is not available, at last count (1985/86) it was 13 TWh (TERI, 1989b). If we project trends for both, we would estimate purchased and self-generated electricity for 1989/90 to be 88 TWh and 15 TWh, respectively, totalling 103 TWh. In Chapter 2 we estimated total 1989/90 electricity sales to be 170 TWh. Adding self-generated electricity, total consumption is 185 TWh. Thus, about 56 per cent of all electricity in India is used by industry.

The major electricity-intensive industries are textile, chemicals, iron and steel, aluminum, fertilizer, cement, paper, nonferrous metals, and collieries. Together they account for over half of industrial electricity use (Anand and Kothari, 1990).

Motors were believed to account for 60 to 70 per cent of electricity use in Indian industry. Until recently, very little was known about the nature of motor use in Indian industry. Anand and Kothari, of TERI, conducted a questionnaire-based survey of industries, followed by field tests and discussions with manufacturers to characterize the electric motor population and conservation potential. Their report, Electric Motors in Industry, submitted to the Indian Government, awaits authorization for public release. The results below are largely taken from a brief article (Anand and Kothari, 1990).

TERI surveyed 11 industries—the nine electricity-intensive industries listed above less collieries and nonferrous metals, plus engineering, jute, sugar, and transport equipment. For these industries, total motor load was estimated to be 13.3 GW in 1983/84. In 1983/84, motors consumed a total of 29.7 TWh in about 70 per cent of the total 42.5 TWh electric consumption. TERI estimates that for all industries, motors consumed 41.8 TWh in 1983/84, or 72.5 per cent of the total 57.1 TWh industrial electricity use. We estimate that industry used 103 TWh of electricity in 1989/90. Thus we would calculate that motors consumed 0.725 x 103, or about 75 TWh.

Squirrel-cage motors account for 69 per cent of motor load and 75 per cent of electricity consumption in the surveyed industries (Table 5.1). Slip ring and other motors are found predominantly in the larger sizes. Anand and Kothari note that, of the total number of squirrel-cage motors, the 9 per cent that exceed 15 kW account for nearly 52 per cent of the total electricity used by motors of this type.

Improving Motor Efficiency

Baldwin (1986), who recently evaluated the opportunities for energy savings in electric motors, noted:

> *During the past twenty years, electric motor technology has sat on the sidelines and watched communication and information technologies explode exponentially in performance. Now, with new developments in power electronics and magnetic materials, electric motor technology itself is being transformed with potentially important implications both nationally and globally.*

Table 5.1
Selected Data on Motors in 11 Industries Surveyed by TERI

Rating	Motor Type			
Range, kW	Squirrel Cage	Slip Ring	Others	Total
Motor Load by Type and Rating, 1983/84, GW				
0.75-3.75	1.35	.03	.021	.40
3.75-7.50	1.04	.04	.03	1.11
7.50-11.25	.99	.07	.02	1.08
11.25-15.00	.94	.04	.01	.99
15.00-37.50	1.50	.26	.07	1.83
37.50-	3.44	1.89	1.59	6.92
Total	**9.26**	**2.33**	**1.74**	**13.33**
Motor Electricity Consumption, 1983/84, TWh				
0.75-3.75	3.31	.04	.06	3.41
3.75-7.50	2.42	.06	.07	2.55
7.50-11.25	2.60	.10	.06	2.76
11.25-15.00	2.50	.07	.03	2.60
15.00-37.50	3.04	.39	.15	3.58
37.50-	8.50	3.38	2.92	14.80
Total	**22.37**	**4.04**	**3.29**	**29.70**

Source: Survey by TERI.

He concluded that more efficient motors could reduce electric demand in new installations for $150/kW and replaced less efficient motors for $525/kW—at a fraction of the capital cost of new power plants. Lovins concluded that the US could save 44 per cent (16 per cent) of the electricity now used in drive systems at an average cost of conserved energy of $0.49 (0.14)/kWh (RMI, 1989).

Energy conservation opportunities in improving motor efficiency may be divided into the following categories:

- Motor type: conventional high-efficiency induction (asynchronous), electronically commutated and AC synchronous permanent magnet, amorphous core, high-speed homopolar, etc.;
- Selection and maintenance: correcting gross oversizing, past rewind core damage, better cooling and cleaning;
- Motor controls: electronic adjustable drives (frequency convertors), slip recovery and hydraulic alternatives, variable speed couplings and mechanical drives, power factor controllers, fast controllers for turbomachinery, idle-off and load management cycling;
- Drivetrain improvements: synchronous and flat belts, improved gearbox designs and types, high-efficiency and magnetic bearings, advanced lubricants, improved lubrication application and maintenance; and
- Electrical tuneups: phase balancing, improved voltage stability and waveform, downstream power factor correction, reducing in-plant distribution losses.

The savings potential depends on the efficiency of existing motors as well as the availability of energy-efficient motors. Anand and Kothari (1990) conducted field tests and compared measured efficiency with rated values. The results of their comparison are summarized in Table 5.2. The measured efficiencies were lower than rated values, partly because Indian manufacturers report design efficiencies rather than those based on measured performance (as in the US), poor motor maintenance, voltage imbalance, age of motor, rewinding, and etc. The discrepancy was larger for smaller-sized motors. Many of the motors were being operated considerably below their rated capacity, with lower actual operating efficiencies (S. Anand, 1990, TERI, pers. com.). The data in Table 5.2 compare full-load efficiencies. This part-load operation was identified as a major element of energy waste in motor operation.

Availability of Energy-Efficient Motors in India

In recent years, considerable advances have been made in the manufacturing capability of energy-efficient motors in India. The Bureau of Indian Standards added a new standard applicable to electric motors. An earlier standard specified a minimum value for the product of the efficiency and the power factor, instead of specifying each one separately. At present the standard applies to motors of the category specified with output up to 37 kW. Later revisions will include other categories and larger motors.

Advertisements for energy-efficient motors appear even in the popular media. At least one major motor manufacturer offers a range of energy-efficient motors. (See Fig. 5.1). The same manufacturer has been promoting a method for saving energy in existing three-phase motors when operated at less than 50 per cent of full load. The motor should be wired in the delta configuration if it is to be operated above 50 per cent full load, and in the star configuration at lower power. The part-load efficiency of the motor is thereby considerably improved (Fig. 5.2).

Table 5.2
Motor Efficiency: Comparison of Manufacturers' Data and Field Test Results

	Efficiency		
Rating, kW	Manufacturers' Data	Field Tests	Ratio of Field Manufacturers' Data
11	87.60	76.90	.88
15	88.60	82.40	.93
18	89.50	82.40	.92
22	90.20	84.60	.94
30	91.50	84.80	.93
37	92.40	86.10	.93
45	93.00	84.60	.91
75	93.70	88.20	.94
11	93.80	89.60	.96
11	93.80	89.60	.96

Source: Anand and Kothari, 1990

Figure 5.1
Nominal Efficiency of Motors

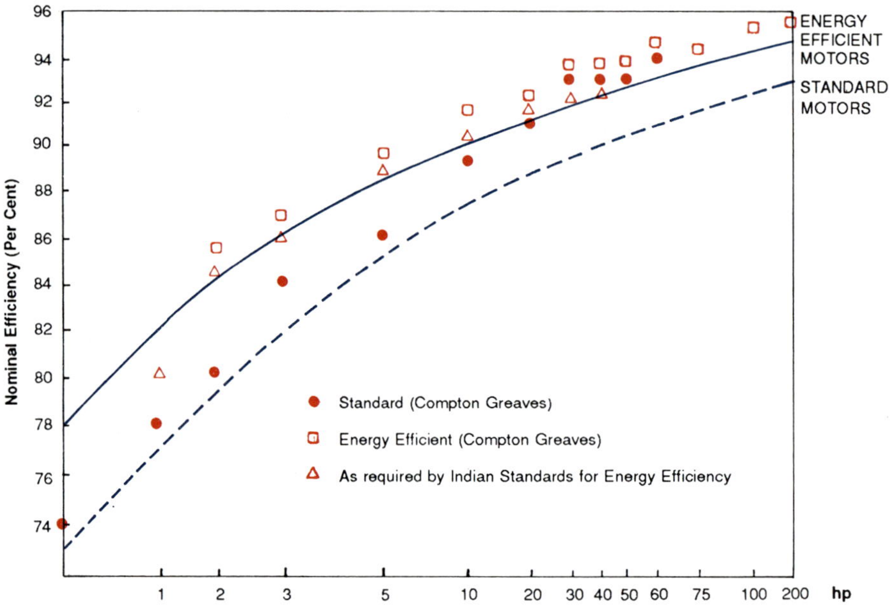

Source : Compton Greaves Ltd.

Figure 5.2

Performance Curves for a 3.7 kW (%HP) 4 Pole Motor with Delta-Star Change Over

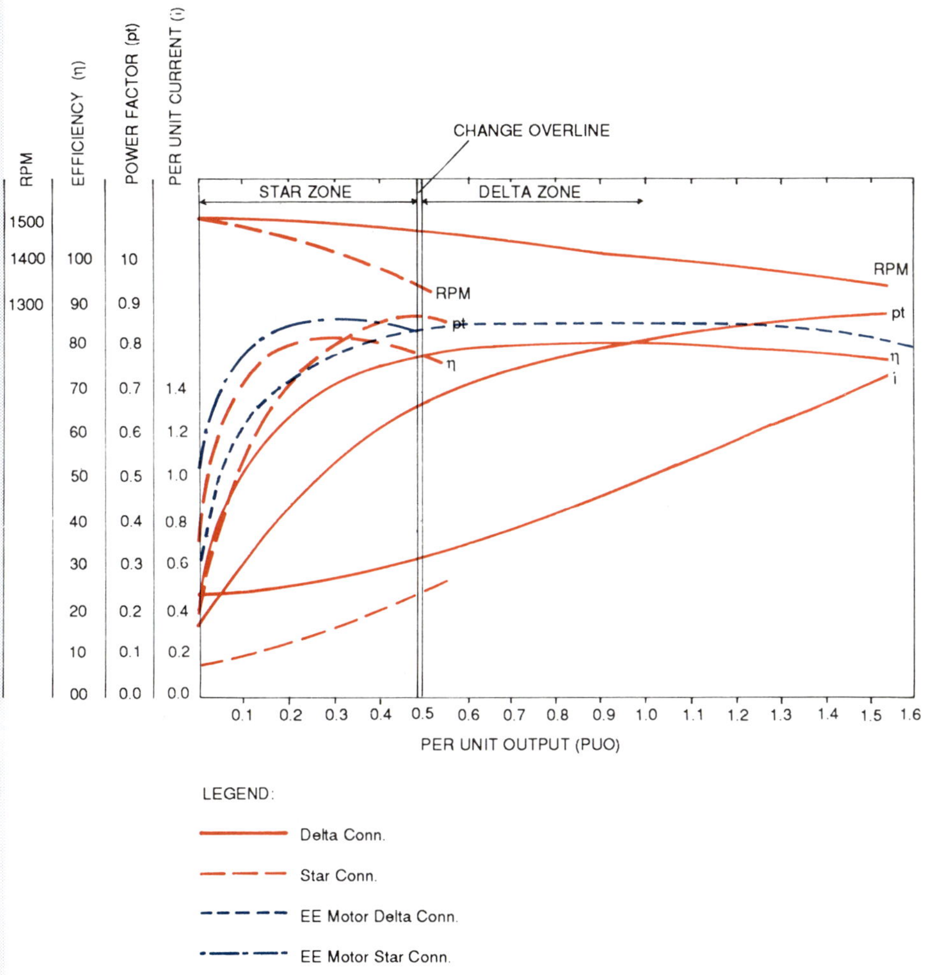

Economics of Energy-Efficient Motors

Let us consider first the case for energy-efficient motors in the US. Gadgil notes that a 10 per cent improvement in efficiency for a fractional horse-power (HP) motor raises the cost by about 30 per cent. For larger motors, the premium is smaller; and for motors of 10 to 200 HP, the extra cost of an energy-efficient motor over a standard motor is approximately $10/HP (Baldwin, 1989). For a 100-HP (output) motor, a 95 per cent motor would cost $70/HP ($7,000), while a 92 per cent one would cost $60/HP ($6,000). The power input would be reduced from 0.746 x 100/0.92 or 81.09 kW to 0.746 x 100/0.95 or 78.53 kW, a reduction of 2.56 kW.

The energy savings and cost-effectiveness depend on the hours of use. For 2,000 hours of use a year, the savings is 5,120 kWh/yr. Assuming a 20-year motor life and a 12 per cent discount rate, the cost of saved energy works out to $0.03/kWh. Clearly, if the motor is used more, the cost of saved energy would decrease. Lovins has calculated the savings with a higher-efficiency motor with the following considerations:

- slightly higher cost/HP;
- lower HP needed by correct sizing; and
- approximately double lifetime between rewinds of high-efficiency motor (14 instead of 7 years).

He notes that one third of motors surveyed by Pacific Gas and Electric Co. (US) were operating at less than 50 per cent of their rated load, and half were operating at less than 60 per cent rated load. As noted above, the field measurements in India also indicate high levels of oversizing. Taking all three factors into account, the immediate replacement of large standard motors with high-efficiency correctly-sized motors yields a cost of saved energy of minus 0.54 cents/kWh (Gadgil, 1990).

Besides energy-efficient new and replacement motors, Gadgil estimated for Indian conditions the cost-effectiveness of a number of other measures for a typical 22-kW motor:

- power factor correction ($0.01/kWh), phase balancing ($0.03/kWh);
- "good" rewinding whereby efficiency is maintained ($0.001/kWh);
- replacing previously-wound motors with efficient (by Indian standards, 91 per cent) motors instead of ones standard (88 per cent) ($0.01/kWh); replacement with international standard (93 per cent) high-efficiency motor, which also

saves $0.01/kWh, but the energy savings potential is much larger); and

- some maintenance measures (at very low cost of saved energy). Gadgil estimates a 16 per cent savings potential from a combination of these measures in the existing stock of motors. This is much more conservative than the 44 per cent potential estimated by Lovins.

Adjustable speed drives, as well as savings in improved pumps, fans, and other elements of the complete system are not considered here.

Energy Savings Potential

Given the present state of information on the condition of industrial motors, the existing full-load efficiencies, the extent of part-load operation, rewinding and maintenance status, and etc., it is difficult to estimate reliably the savings potential. Assuming that all new motors added during the 1990s are up to the Indian or international standards for high efficiency, Anand and Kothari (1990) estimate the savings potential to be 0.25 TWh/yr or 0.58 TWh/yr, respectively. Using Gadgil's estimate of 16 per cent savings in the electricity used in existing motors, we would add another 0.16 x 75 or 12 TWh a year for Indian industrial motors in 1989/90. Except for phase balancing, all measures have a cost of saved energy of $0.01/kWh or less.

To estimate a savings potential we first estimate the year 2000 if there were no efficiency improvements, i.e., a "frozen-efficiency" consumption. We assume a 25 year motor life and a 7.4 per cent annual growth rate in motor load (Anand and Kothari, 1990). Some of the current stock of motors would survive till the year 2000, and new motors will be installed to meet future demand growth. If all use the same amount of energy, total consumption in the year 2000 will be 1.07411 x present (year 1989-90) consumption = 2.19 x 75 TWh = 164.5 TWh. To estimate the savings potential we need to make additional assumptions. We assume that half of the existing (1989-90) stock of motors would survive until the year 2000. Gadgil (1990, personal communication) has estimated a 16 per cent savings potential in existing Indian motors through a combination of power factor correction, phase balancing, good rewinding, or replacing previously wound motors with better-sized, energy-efficient motors, by Indian or US standards. (Most of these measures have estimated costs of saved energy of around $0.01 per kWh, far lower than the marginal cost of electricity supply [around $0.09 per kWh], indeed far lower than industrial sector electricity tariffs in India of around $0.05/kWh.) We assume that new installations would save somewhat more, i.e. consume 20% less than the frozen-efficiency case. (Note that Lovins estimates that the US could save approximately 44 per cent of the electricity now in use in drive systems at an average cost of conserved energy of $0.49 per kWh [RMI, 1989].) With these assumptions, the energy consumption of industrial motors in the year 2000 conservation scenario is as shown in Table 5.3.

Table 5.3
Year 2000 Electricity Consumption of Industrial Electric Motors (assuming efficiency improvement)

Category of Motor	Year 2000 consumption (TWh)		
Existing motors, surviving to 2000	0.5 x 75 x (1- 0.16)	=	31.5
Existing motors, replaced by 2000	0.5 x 75 x (1- 0.20)	=	30
New installations, 1990-2000	$(1.074^{11} - 1)$ x 75 x (1- 0.20)	=	71.6
Total consumption with conservation, year 2000		=	**133.1**

Source: Anand and Kothari, 1990.

Thus the electricity conservation potential, compared to the fixed efficiency scenario, in the year 2000, is 31.4 TWh.

(Due to lack of data, we have not been able to estimate savings potential through variable-speed drives, nor in other components of the motor drive system, such as pumps, blowers, piping, etc. These will considerably increase the savings potential.)

Policy Measures

Despite the large potential for energy conservation and relative cost-effectiveness, the penetration of energy efficiency into the industry is likely to be slow. Given the enormous benefit to society, many steps may be taken to facilitate the conversion.

New energy-efficient motors. Anand and Kothari (1990) made several recommendations to promote energy-efficient industrial motors. At the time of their analysis (completed in 1988), there were no standards for energy-efficient motors and very little (if any) domestic manufacturing capability. We have adopted their recommendations and added some others.

We believe that the availability of inefficient motors should be phased out as soon as possible. The manufacturers' associations and the Government should agree on a timetable for bringing the average motors sold up to the Indian Standard energy efficiency level within say, four years, and to international levels in, say, seven years. The supply of necessary raw materials to the manufacturers should be streamlined. Financing for upgrading manufacturing facilities should be given priority. The national industrial development banks, Industrial Development Bank of India (IDBI), Industrial Credit and Investment Corporation of India (ICICI), and other lending institutions could play an important part in this. Technical assistance could be made available to small-scale manufacturers. Courses organized by the National Productivity Council, TERI, Energy Management Center, Indian Electrical and Electronic Manufacturers' Association, and others should be financially supported. (The Indian Electrical and Electronic Manufacturers' Association includes the larger companies; All-India Electric Motors Manufacturers' Association includes mainly the smaller manufacturers.)

Testing and verification. Additional testing facilities, particularly in the private sector, should be set up to evaluate the performance of new motors. Motor efficiency ratings and other characteristics should be measured. The current practice of using design or "nominal" performance figures should be discontinued as soon as possible.

Users need some incentive to purchase energy-efficient equipment. At present, excise duties are levied as a fixed percentage of the price of the product. Reductions in excise duty for energy-efficient systems have been proposed. This may be complicated and difficult to administer. One way to eliminate the current "penalty" for the purchase of an efficient motor (that is somewhat more expensive) would be to base the excise duty on the input kW of the motor, rather than on the price. This will also provide an added incentive for downsizing, whenever possible. Measures directed at improving the efficiency of motors will affect motors used in other sectors as well as industry.

Existing motors. Achieving savings will be more tedious here, since it will depend critically on the availability of a large number of professionals trained to maintain, diagnose, repair, and rewind motors. Apart from training, motor auditors will need appropriate diagnostic equipment, and possibly inexpensive computers to calculate motor performance from measured data (see, for instance,

Baldwin, 1988). Compared to the immensity of the task, relatively few people in India are currently qualified to carry out diagnostics. The 11 industries surveyed by TERI were estimated to have over a million motors in 1983/84 (Anand and Kothari, 1990). Training may be facilitated through courses with the experts from industrialized countries. Japan International Cooperation Agency or the European Economic Community (EEC) could support such efforts, focusing on the training of local trainers. The Energy Management Center is ideally suited to organize this task as part of its Energy Bus Program which has mobile industrial energy audit units financed by the EEC and by the United Nations Development Programme (EMC, 1990).

Improved data. Considering that over 40 per cent of Indian electricity is used in industrial motors, very little information is available to characterize them. The agricultural pumping retrofits (Chapter 4) illustrate the tremendous conservation opportunities available in motor drive systems.

Many industrial energy audits and energy conservation projects have been carried out in India by both government and private groups. When reports are available, they tend to summarize overall savings potential and/or achievements, without breaking them down by end uses. A data base of estimated and actual savings by end uses needs to be set up so that better alternatives may be promoted and less useful measures discouraged.

A technology menu for energy-efficient motor drive systems, currently under preparation by Princeton University's Center for Energy and Environmental Studies, should help bridge the information gap in both domestic manufacturers and potential users.

6

Comparative Costs of Electricity Conservation

In Appendix I (p. 196) we present a methodology for evaluating the cost-effectiveness of energy conservation opportunities using two indices—the cost of saved energy, and the annualized life-cycle cost. Typically, the cost of saved energy is used to compare conservation measures against one another and with the unit price of energy. For electricity, the price of energy is determined by the costs of operating the mix of electric power plants that provide electricity, plus any taxes and subsidies that may be in effect. A large part of the material in this chapter is adapted from the work of A.K.N. Reddy, G. D. Sumithra, P. Balachandra, and A. d'Sa, one version of which appeared in Economic and Political Weekly (India), 2 June 1990.

Reddy et al. extended the concept to permit a comparison of electricity conservation opportunities against an extrinsic electricity price, and a comparison of alternative generation technologies with each other and with the conservation options. They used this methodology, with empirical data from Karnataka state and elsewhere in India, to compare a number of centralized and decentralized electricity generation options and some electricity conservation measures. The calculation of the cost of electricity generation is described below.

Electricity Generation Economics

The method is based on calculating the Present Value of all the costs incurred in constructing and operating a power plant during a fixed time interval. Many power plants require expenditures spread over several years, and the amounts for each year are converted to an equivalent amount for a given reference year, using a discount rate. Since the various power plants (and conservation measures) have different lifetimes, all costs are computed over a 25-year period, corresponding to the options with the longest life. Repeated investments of shorter-lived items are deemed necessary to reach the 25-year span. This Present Value of all costs is divided by the power generated (kW) and the total energy generated (kWh) to determine a unit cost for power and energy. Due to transmission and distribution losses, much more than a unit of electricity (power or energy) needs to be generated in central station power plants in order to deliver a unit to the point of end use. Thus, the cost of electricity generated in centralized plants was adjusted to determine an equivalent cost per kW or kWh delivered. Decentralized electricity generation and electricity conservation options do not need to be adjusted for transmission and distribution losses. The resulting unit costs of power or electricity delivered to the end use or saved at the end use can then be compared in order to determine the relative economics of the options being considered.

The cost assumptions used for the various generation options are given below. Reddy gives the cost figures in 1986 rupees. Assuming a 7 per cent per year inflation rate, one 1986 rupee would equal Rs 1.31 in 1990. We then convert this to 1990 US dollars using the 1990 year-end exchange rate of Rs 18/$. The net result is an effective exchange rate of Rs 13.73 (1986) = $1 (1990). We used this factor to convert Reddy's cost figures to 1990 $. In the following discussion, unless otherwise stated, we use 1986 Rs as in the Reddy analysis and equivalent 1990 US dollar values.

Environmental costs were not considered in this analysis. One reason for not including environmental consequences in terms of shadow costs is that these tend to be much more controversial than the actual costs of construction and operation. A comparison among the alternatives based on a consistent set of economic assumptions (notably the discount rate) using a methodology that permits a rational comparison favors many of the more environmentally benign technologies, even without including any environmental costs. Adding environmental considerations will tilt the balance further in this direction. For a brief discussion on how environmental considerations may be quantitatively included in the analysis, please refer to the Environmental Costs section of this chapter.

Nuclear power plants (CANDU type). The total annual undiscounted expenditures for the construction of two 235-MW plants for completion in 1992 is Rs 7,240 million. The expenditure stream may be discounted to a present value using a discount rate of 12 per cent, giving an "overnight" capital cost for construction of Rs 6,080 million (1990 $443 million). The "overnight" capital cost per unit of power generated is Rs 12,941/kW ($943/kW). These cost figures are based on a report of India's Department of Atomic Energy (DAE, 1985; M. R. Srinivasan, 1985; Srinivasan and Mahadeva, 1987). Williams and Larson (1989) cite the Electric Power Research Institute estimate for the capital cost of a 1,100-MW light water nuclear reactor in the US to be $1,100/kW (1987), or about $1,237/kW (1990), assuming a 4 per cent inflation rate. Reddy considers four other cases using other data for the cost of the heavy water moderator (from another government department [CAG, 1988] the cost of waste disposal, based on Swedish experience, is Rs 0.05/kWh or $0.0028, and 10 instead of 8 years for constructing the reactors). The Atomic Energy Commission's cost estimate for plant decommissioning, Rs 1,000/kW ($73/kW), was used even though this is believed to be unrealistically low. Using additional official data on fuel prices, efficiency, operating costs, and a capacity factor of 62.8 per cent, the cost of power generation was calculated per unit power (kW) and per unit of electrical energy (kWh). These costs are adjusted to delivered, or end-use, power and energy using transmission and distribution losses of 22.5 per cent. At a discount rate of 12 per cent, the range of costs varies from Rs 73,704 to 108,216/kW, and Rs 1.85 to Rs 2.74/kWh for the five cases considered by Reddy ($5,368 - 7,882/kW, and $0.13 - 0.20/kWh).

Coal-based power plants. Reddy used the same data source for coal as for the CANDU nuclear reactor. The capital cost sum for a pair of 210-MW plants is Rs 6,210 million. At a 12 per cent discount rate, the overnight capital costs are Rs (1986) 4,690 million ($342 million) or Rs 11,174/kW ($814/kW). Again, this may be compared with an Electric Power Research Institute (EPRI) estimate for a 200-MW coal-fired power plant with flue gas desulfurization in the US of $1,880/kW in 1987, or $2,115/kW in 1990 (Williams and Larson, 1989). Two cases were considered, a power plant located at a coal mine, and one located 800 km away, where coal transportation is necessary. Again, using official data on fuel prices, efficiency, operating costs, and a capacity factor of 62.8 per cent, cost of delivered power and energy were calculated. At a discount rate of 12 per cent, these are Rs 53,400 - 56,959/kW, and Rs 1.74 - 1.83/kWh ($3,889 - 4,149/kW, and $0.13/kWh).

Hydroelectric plants. The capital cost requirements of hydroelectric plants tend to be very site-specific. Moreover, the environmental costs can be large and perhaps immeasurable. The environmental costs may be difficult to quantify, especially:

- loss of biodiversity, forest cover and agricultural land; and
- displaced populations, including indigenous peoples whose entire habitat may be flooded, leading to cultural extinction.

Large hydroelectric plants have come under increasing opposition even in developing countries. Emotions run high in India over the Narmada project, a series of dams for providing irrigation and electricity to Western Indian states. On one side are the contractors who will get hundreds of millions to billions of dollars for the construction of dams and associated infrastructure. On the other side are the environmentalists and local populations taking desperate stands to prevent the project.

We must keep these factors in mind as we look at the example in which Reddy considers the monetary costs of a specific hydroelectric project—Varahi in Karnataka. The undiscounted sum of capital expenditures is Rs 2,100 million for this 230 MW plant, commissioned in 1989 (KPC, 1985-86). At a 12 per cent discount rate, the overnight capital costs are Rs 2,384 million ($174 million) or Rs 10,366/kW ($755/kW). For a capacity factor of 65 per cent and estimated annual operating costs of 1 per cent of capital, the cost of delivered energy and power are Rs 54,538/kW and Rs 1.25/kWh ($3,972/kW and $0.09/kWh).

Natural gas-fired power plants. Natural gas-fired power plants present another option for centralized power generation. The total nominal expenditure of Rs 5,690 million for a 760-MW plant due for completion in 1989 converts to an overnight construction cost of Rs 4,960 million ($361 million) or Rs 6,527/kW ($475/kWh). The corresponding US value for a 2 x 500-MW plant based on steam turbines is $760/kW (1987) or $855/kW (1990), and for a 205-MW advanced combined cycle plant is $520/kW (1987) or $585/kW (1990) (Williams and Larson, 1989). The Power Sector Group of the ADB informally estimated this cost to be somewhat higher—about $700/kW (1990), based on plants being considered for financing by the Bank. Using the Indian data, Reddy et al. (1990) obtained the following results: at a gas price of Rs 1,400 ($102) per standard cubic meter, assumed operating and gas transportation costs, and a capacity factor of 62.8 per cent, the costs of delivered power and energy are Rs 44,054-50,115/kW and Rs 1.08 - 1.24/kWh ($3,209 - 3,650/kW and $0.08 -0.09/kWh) for the two cases of plant location—at the source of natural gas and 800 km away.

Decentralized electricity generation. Besides the three "conventional" centralized technologies (coal, nuclear, and large hydroelectric), and the relatively new (to India) natural gas option, Reddy also considered several decentralized electricity generation options, both renewable and nonrenewable. Among the renewable options are small hydroelectric plants, biogas- and producer gas-based power plants, and wind farms. Diesel engines are an alternative based on nonrenewable fossil fuels. Since decentralized plants are intended to be near the site of power demand, Reddy considers no transmission and distribution losses in converting from generated to delivered power. In reality there will be some losses in local distribution, and perhaps in voltage transformation as well.

Small hydroelectric plants. The capital costs of small hydroelectric plants are also likely to be site-specific. Reddy considers data from a mini-hydroelectric project, again in Karnataka (KPC, 1982). The entire capital cost of Rs 106 million for this 9 MW project has been considered to have occurred at the beginning of the project in 1982 (thereby increasing the amount). In 1986, this would be equivalent to Rs 15,395/kW ($1,121/kWh). With additional assumptions, the costs of generated and delivered power and energy were calculated to be Rs 16,644/kW and Rs 0.43/kWh ($1,212/kW and $0.03/kWh).

Biogas plants. The cost figures for generating electricity are based on a study at a community biogas plant at Pura Village in Karnataka. Biogas produced in two plants by the anaerobic fermentation of cattle dung is used to run a 7-P dual fuel engine that runs on 80 per cent biogas and 20 per cent diesel and operates a 5-VA generator. Since this type of system is unfamiliar, particularly outside India, we give more details than in the alternatives discussed above.

The total engine life is assumed to be 20,000 hours, including three overhaulings. The capital costs of the different elements of the complete system are given in Table 6.1. Assuming a capacity factor of 62.8 per cent and a discount rate of 12 per cent, the costs of power and energy are Rs 47,855/kW and Rs 1.23/kWh ($3,485/kW and $0.09/kWh).

Table 6.1
Capital Cost Details of Biogas-Based Electricity Generation

Item	Capital Cost[a]	
	Rs (1986)	$ (1990)
Biogas plant	28,794	2,097
Piping, etc.	1,869	136
Sand filters	935	68
Engine & generator	26,122	1,903
7-HP diesel engine	8,500	(619)
5-kVA 3-phase generator	17,622	(1,283)
Tools and accessories	4,673	340
Engine room	3,738	272

a Actual costs adjusted to 1986 levels using an inflation rate of 7 per cent per year. Conversion to 1990 is based on a conversion factor of Rs 13.73 (1986) = $1 (1990) as described in the text.
Source: Reddy et al., 1990

Most biogas plants in India (over a million by 1988 [Dayal, 1989]) are of the individual household capacity which produces too little gas to operate an engine generator. By the end of 1988, 430 community and institutional biogas plants had been installed in India (TERI, 1989b). Few are used to generate electricity.

In a comparison of alternative strategies for providing motive power to a farm, Bhatia (1990) compared four options—electricity from the grid, diesel engines, biogas and diesel, and producer gas and diesel—using both shadow prices and actual prices. He found the biogas/diesel to be the least expensive option using shadow prices (i.e., from a societal perspective), but because of subsidies, electricity from the grid was the least-cost alternative from the farmer's perspective.

The environmental costs of biogas plants are expected to be negative. The digester results in higher nitrogen content in the dung than if it had decomposed aerobically. Moreover, no net carbon dioxide emissions result from the conversion process. (The same amount of carbon dioxide is produced by aerobic decay and by combustion following biodigestion.)

Producer gas plants. The data for producer gas plants are based on the operational experience in Hosahalli, a Karnataka village.

A wood gasifier is used in a 7-HP dual fuel engine coupled to a 5 kVA generator, as in the Pura biogas system. The capital costs are thus the same as in the biogas case, except for the producer gas plant and a wood-chipping machine (Table 6.2). For a capacity factor of 62.8 per cent, the cost of power and energy are Rs 51,646/kW and Rs 1.33/kWh ($3,762/kW and $0.10/kWh).

Table 6.2
Capital Cost Details of Producer Gas-based Electricity Generation

Item	Capital Cost[a]	
	Rs (1986)	$ (1990)
Producer gas plant	16,500	1,202
Voltage stabilizer	1,500	109
Water pump	1,500	109
Wood chipper	3,000	218
Engine & generator	26,122	1,903
7-HP diesel engine	8,500	(619)
5-kVA 3-phase generator	17,622	(1,283)
Tools and accessories	4,673	340
Engine room	3,738	272

a Actual costs adjusted to 1986 levels using an inflation rate of 7 per cent per year. Conversion to 1990 is based on a conversion factor of Rs 13.73 (1986) = $1 (1990) as described in the text.

Source: Reddy et al, 1990

At year-end 1990, there were over 300 gasifiers in operation in India, including 120 in Karnataka of the design used in Reddy's analysis (Muthusubramanian and Haridasan, 1990). The Gasification Research Center of the Madurai Kamaraj University evaluated the performance of two gasifier designs through testing. Testing, lasting 1,100 hours in one case and 450 hours in the other, revealed a few minor materials problems that are correctable by using more resistant materials in critical areas such as the gasifier throat. The dust filtering was found inadequate in one design. Diesel substitution rates of 60-70 per cent and 70-80 per cent were observed in the two designs.

Wind farms. Reddy's analysis is based on the wind farm at Okhla in Gujarat State (GEDA/GEB, n.d.). An investment of Rs 8.6 million in 1986 for a generation capacity of 550 kW is equivalent to an overnight construction cost of Rs 17,899/kW ($1,304/kW). Since generation depends on wind speed, the capacity factor is site-specific. In this case it was no more than about 20 per cent. The costs of power and energy are found to be Rs 40,061/kW and Rs 1.96/kWh ($2,918/kW and $0.14/kWh).

Diesel. The diesel-generation costs have the same basis as biogas and producer gas except that no gas generator is needed and the engine works on a single fuel. The Rs 34,533 capital cost for a 5-kW plant is equivalent to an overnight cost of Rs 6,907/kW ($503/kW). For a diesel price of Rs 4.45 ($0.32) per liter, unit costs of power and energy were calculated to be Rs 60,481/kW and Rs 1.56/kWh ($4,405/kW and $0.11/kWh).

Cogeneration in sugar factories. Reddy considered one case of cogeneration as an electricity supply option. Sugarcane bagasse is sometimes burnt to meet the steam requirements of sugar factories. A part of the steam is also used to generate electricity for the mill. Traditionally, in order to avoid a bagasse disposal problem, sugar factories were designed to consume all of their bagasse to meet plant steam and electricity requirements. Recently, an alternative approach has been proposed—to use the bagasse to generate electricity using efficient conversion technologies that yield a surplus of electricity beyond the needs of the factory. This excess can be "exported" as another product of the sugar mill (Larson et al., 1989).

Prabhakara Rao (1989) conducted a cogeneration feasibility study at a sugar factory in Karnataka. Reddy considered two cases:

- where the plant operates only during the cane milling season; and

- where the plant operates in the off-season as well, using coal fuel.

In either case, the overnight construction cost was estimated at Rs 13,500/kW ($983/kW). For bagasse only operation, the capacity factor of the plant was estimated to be only 31.7 per cent, while with bagasse plus coal, it rises to 52.2 per cent. At a 12 per cent discount rate, the unit cost of power and energy is calculated to be:

- bagasse only—Rs 28,138/kW ($2,049/kW) and Rs 0.64/kWh ($0.05/kWh)
- bagasse + coal—Rs 20,899/kW ($1,522/kW) and Rs 0.47/kWh ($0.03/kWh).

Reddy treats cogeneration as a decentralized option and does not adjust for transport and distribution losses. However, some adjustments may be in order for sugar factories located away from load centers.

The world sugarcane resource is very significant. The 1987 world production of sugarcane was almost a billion tons, and production was growing at about 3 per cent per year. The global average cane productivity is about 58 t/ha per year, equivalent to a woody biomass production of 38 dry t/ha per year. This is substantially higher than can be realized on any energy plantation (Ogden, Williams and Fulmer, 1990). The current and future cane production and electricity generation potential using efficient conversion processes are shown in Table 6.3. In India, the potential for sugarcane-generated electricity in the year 2027 is 2.6 times larger than the 1987 generation of electricity from all other sources, with comparable ratios for many other Asian countries with significant cane production.

A major environmental attraction of the bagasse cogeneration route is that, except to the extent that coal is used in the off-season, it is a renewable option which adds no net carbon dioxide to the atmosphere.

Cost of Saved Electricity

Reddy's comparison of the unit costs of power and energy using the various alternatives discussed above is presented in Figures 6.1 and 6.2. These figures include the cost of saved electricity from several conservation measures—solar water heaters, improved irrigation pump sets, replacement of standard motors with efficient motors, replacement of incandescent lamps with compact fluorescent lamps, and solar ponds. The assumptions made in these measures are discussed below.

Solar water heaters. The cost of solar water heaters in 1988 was Rs 3,000, but Reddy et al. (1990) assumed a 1986 cost of Rs 5,000 ($364), allowing for some system improvements. A savings of 577 kWh/yr, equal to the average electricity use of a conventional electric water heater, is assumed (except for about 35 days a year when the solar water heater cannot be used). Unit costs of (saved) power and energy are Rs 2,834/kW ($206/kW) and Rs 1.58/kWh ($0.12/kWh). While the unit cost of power is exceptionally low, the energy cost is not, reflecting the low hours of usage—a conventional water heater is used only some 365 hours a year (Sudhakar Reddy, 1990).

Table 6.3
Electricity Production Potential in the Sugarcane Industries of Developing Countries

	A	B	C	D	E = C/D
	Cane 1987[a]	Production 2027[b]	Electricity Potential from Cane in 2027[c]	Production Actual from All Sources in 1987[d]	
	Million t of cane/Yr		TWh per Year		
India	182.48	629.55	557.15	217.50	2.6
China	52.55	181.30	160.45	497.30	0.3
Pakistan	31.70	109.37	96.80	28.40	3.4
Thailand	24.45	84.35	74.65	29.99	2.5
Indonesia	21.76	75.09	66.45	34.81	1.9
Philippines	13.33	45.97	40.68	23.85	1.7
Bangladesh	6.90	23.79	21.06	5.90	3.6
Viet Nam	6.60	22.77	20.15	5.20	3.9
Burma	3.28	11.32	10.02	2.28	4.4
Iran	1.15	3.97	3.51	36.80	0.1
Malaysia	1.15	3.97	3.51	16.22	0.2
Sri Lanka	0.80	2.76	2.44	2.71	0.9
Nepal	0.62	2.13	1.88	0.54	3.5
Kampuchea	0.21	0.71	0.63	–	–
Laos	0.11	0.36	0.32	0.88	0.4
Subtotals	347.10	197.40	1,059.70	902.40	1.2
Totals	**911.50**	**3,144.40**	**2,782.80**	**1,632.40**	**1.7**

a FAO, 1987.
b Assuming sugarcane grows at a rate of 3.1 per cent/yr.
c For sugar factories or alcohol distilleries operated 133 days/year, with 286 kWh of electricity produced from bagasse/ton of cane during the milling season plus 599 kWh produced from bagasse/tc during the off-season.
d Except where otherwise indicated, from Escay, 1990.
e For 1986, from *Statistical Abstract of the United States 1989*, 1989.
f Public electricity production in 1989. From Moore and Smith, 1990.
Source: Ogden et.al. 1990.

Improved irrigation pump sets. Only two simple improvements have been considered for irrigation pump sets (IPS)—a frictionless foot valve costing about Rs 100 each and saving 215 kWh/IPS per year; and high-density polyethylene piping instead of galvanized iron, costing about Rs 1,000/IPS and saving about 495 kWh/IPS per year. The cost of saved power and energy are Rs 840/kW ($61/kW) and Rs 0.27/kWh ($0.02/kWh). While this is clearly a bargain, the Gujarat experience (see wind farms, p. 65) shows larger savings potential at a lower cost of saved electricity—1,400 kWh/yr at $0.01/kWh.

Energy-efficient motors. Capital costs of standard and improved 5-HP motors are assumed to be Rs 3,550 and 4,550, with efficiencies of 82 and 87 per cent, respectively. An annual usage of 3,000 hours and a motor life of 20 years are assumed. The cost of saved power and energy are calculated to be Rs 2,185/kW ($159/kW) and Rs 0.15/kWh ($0.01/kWh). Note from the discussion in Chapter 5 that the average industrial motor is much larger, but hours of use per year are less than values assumed here. Irrigation pump sets are this size but usage is typically less than 1,000 h/yr (Ramesh and Natarajan, 1987).

Compact fluorescent for incandescent lamps. Reddy assumes a 60-watt, 695-lumen, 1,100-hour incandescent bulb, costing Rs 5.60 ($0.41) will be replaced by a 9-watt, 700-lumen, 6,000-hour compact fluorescent lamp costing Rs 106 ($7.72). The separate base with ballast for the compact fluorescent lamp costs Rs 116 ($8.45), consumes 6 watts, and lasts 10 years. Each lamp is assumed to be used 1,825 hours per year (5 hours a night). The cost of saved power and energy is Rs 4,395/kW ($320/kW) and Rs 0.49/kWh ($0.04/kWh).

Solar ponds. Lastly, Reddy considered solar ponds—covered ponds used to receive and store solar energy in the form of hot water. A 250-square meter pond is estimated to cost Rs 120,000 ($8,740) and result in saving 50,000 kWh/yr. The functioning of the pond is described by J. Srinivasan (1985). The costs of saved power and electrical energy are calculated to be Rs 15,033/kW ($1,095/kW) and Rs 0.35/kWh ($0.03/kWh).

The five saving options. The comparison of these conservation and generation alternatives, at a 12 per cent discount rate, per unit power and energy (Figures 6.1 and 6.2) reveals that the five conservation options are the lowest cost options on a unit power basis. The ordering, per unit of electrical energy, is somewhat different, but three of the conservation options still come out cheaper. Reddy notes with irony that the conventional power generation alternatives are among the most expensive. In Chapter 7 we consider how these relative costs may be used to define a least-cost strategy for providing electricity. Below, we make some additional comments related to the analysis of this chapter.

Peak and Base Load

The above analysis compares generation and conservation alternatives without attention to the timing of the demand. Since electricity cannot be economically stored in significant amounts, generation must always match demand. The variation in demand is characterized by a load duration curve which tells us what fraction of the time a certain level of demand is exceeded. The level of power demand that exists virtually all the time is met by base load plants, while increasing levels of demand are met by load-following and peaking power plants. Since power plants vary in their capital and operating costs, the most economical base load plant is usually different from a load-following or a peaking power plant. A standard, transparent method for determining the optimal mix of power plants uses the so-called screening curves. For an excellent discussion with examples of how this works, see ABB (1989). The screening curves can also be used to calculate the cost of power generation using the different power plants, given their capacity factors as needed to meet the demand. The cost of electricity generated in the peaking plants is higher than in the base load plants.

A simple method for taking the load duration curve and optimal power plant mix into account in comparing conservation and generation alternatives does not yet exist. The above method considers basically a number of base load power plants and compares them with a number of conservation options. For only one of the conservation measures—light bulbs—the above analysis tends to be somewhat biased. Incandescent lamps, particularly in the residential sector, are used in the evenings when they contribute significantly to the peak demand, which must be met at a higher cost (assuming that the peak can be met, which is not currently the case in India). Thus the cost-effectiveness of the lamp retrofit is better than would appear in the above comparison.

Electricity Generation Using Biogas or Producer Gas Dual-Fuel Engines

The analysis above shows that the biogas or producer gas dual-fuel options produce electricity at $0.09 to 0.10/kWh, which is less than centralized production from coal, nuclear or large hydroelectric plants.

Figure 6.1
Unit Cost of Power
(Including Transport and Distribution for Centralized Sources)

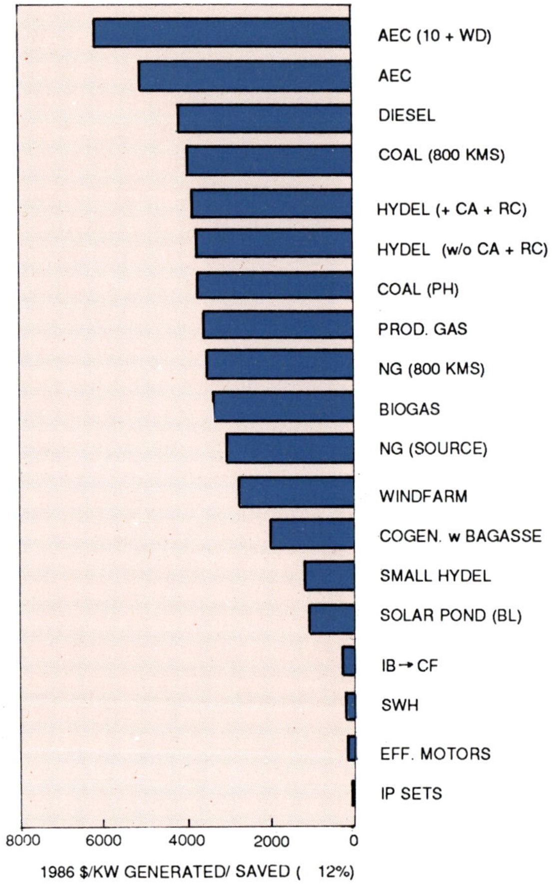

Source : Reddy et. al., 1990

Figure 6.2
Unit Cost of Energy
(Including Transport and Distribution for Centralized Sources)

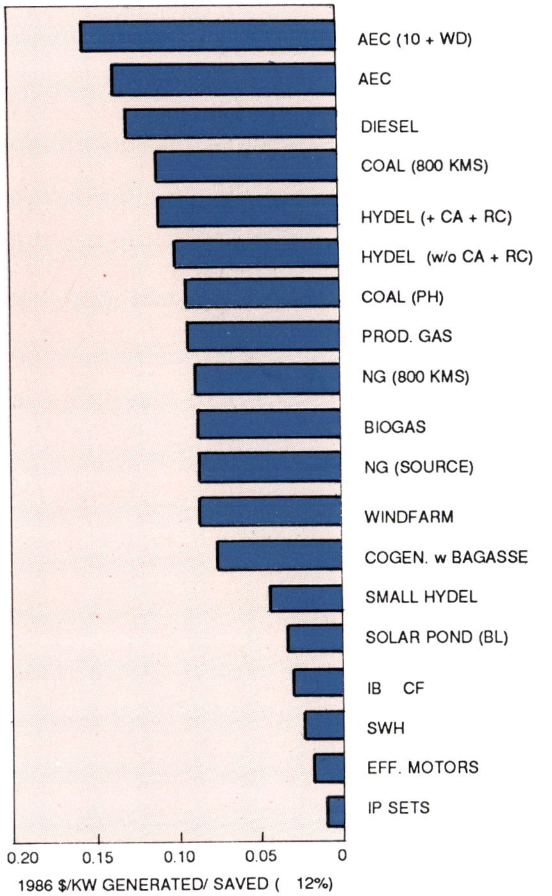

Alternative Fig. 7.2 in 1986 $
Source : Reddy et. al. 1990

This analysis however was based on an assumed 62.8 per cent capacity factor, i.e., 15.1 hours of operation. It may not be possible to find such an extended demand in rural areas where the principal uses are likely to be lighting and water pumping. In Pura Village, the source of Reddy's data for the biogas costs, engine usage from January 1988 to March 1990 averaged less than four hours a day.

Such a low load factor would favor the diesel engine, with capital cost a little over a half that of the biogas dual-fuel system. We noted in Chapter 2 that where households use kerosene for lighting, an eightfold fuel savings would result if diesel generators lit fluorescent lamps, instead of using kerosene lamps. Since kerosene is subsidized more than diesel, social savings would be significant while the households' cost of lighting would be greatly reduced as well.

Another way to deal with the low load factor problem would be to use biogas or producer gas engines for water pumping instead of generating electricity. The cost of generating equipment is a significant part of the total cost of biogas or producer gas-based electric plants (see Table 6.1). Since water can be pumped with dual-fuel diesel pumps as well as by electric pumps, this option would be economical. Later, as electric demand increases, an electric generator may be added.

Environmental Costs

The environmental consequences of energy projects are quite significant. They include:

- local effects such as air pollution (nitrogen oxides, sulfur oxides, particulates);
- medium-distance phenomena such as acid rain; and
- long-range, long-term climate change such as global warming from the emission of carbon dioxide and other "greenhouse gases" and depletion of the stratospheric ozone layer from the emission of chlorofluorocarbons and other ozone-depleting substances.

Coal power plants generate local air pollution, acid rain, and carbon dioxide. Local air pollution and gas emissions that lead to acid deposition can be reduced greatly through appropriate design changes. These are called "clean-coal" technologies, and have become standard in industrialized countries. Carbon dioxide emissions per unit of electricity generated can be reduced using technologies that permit a higher efficiency of conversion from coal to electricity. Alternatively, the carbon dioxide can be "sequestered," e.g., the amount of carbon dioxide added to the atmosphere in a

500-MW power plant operating at 34 per cent efficiency may be removed by planting a forest of about 1,000 square miles (Fulkerson, Judkins and Sanghvi, 1990).

Nuclear power plants do not lead to local air pollution, acid rain, or global climate change. The principal environmental consequences are related to the handling of nuclear materials. Since atomic bombs can be readily made from nuclear fuels, there is an increased risk from proliferation of nuclear weapons by national governments and by terrorist groups, as more nuclear material circulates in the world economy. Another risk is in the disposal of radioactive waste. Since some of the products of a nuclear reactor are radioisotopes with half-lives of thousands of years, they need to be stored in geologically stable locations. Moreover, many people are concerned about the catastrophic consequence of a nuclear accident that can radioactively contaminate large areas for long periods of time.

Some environmental costs may be internalized by using cleaner or safer technologies. Alternative estimates that take some of the environmental costs into account are the so-called clean-coal technologies typical of industrialized countries. Table 6.4 compares a number of coal and natural gas-fired electricity generation technologies on the basis of cost, efficiency, and emission of sulfur dioxide, nitrogen oxides, and carbon as carbon dioxide. The principal conclusion is that alternative technologies based on coal can bring down costs and substantially reduce emissions of sulfur dioxide, nitrogen oxides, and carbon dioxide, compared to those of the conventional steam turbine option.

Another conclusion from Table 6.4 is that gas-fueled technologies are far less capital-intensive, and they are cleaner than any of the clean-coal options. Natural gas supplies are very unevenly distributed, however, and pipeline construction can be capital-intensive. For instance, a 2,000-mile pipeline from Bandar Abbas (Iran) to Calcutta, providing gas to Pakistan and India, is estimated to cost $12 billion. Natural gas-fired combined-cycle plants have been set up in a number of developing countries, and more are under construction. Table 6.4 shows that steam-injected gas turbines (already commercialized) and an intercooled version capable of higher efficiency may turn out to be less expensive than combined-cycle plants. One important advantage of gas turbines over steam turbines (which are a part of combined-cycle plants) is that economies of scale are weak so that economically viable plants can be much smaller, perhaps at a scale more suitable for developing countries. Smaller plants also can be built in less time, adding considerable flexibility to power system planning (Williams and Larson, 1989).

Table 6.4
A Comparison of Alternative Technologies for Power Generation from Natural Gas and Coal

	Plant Cost $/ kW	Efficiency per cent	Reduction in SO_2 Emission % mg/ kWh	Emission of NO_x kg/ kWh	Emission of Carbon kg/ kWh
Conventional (steam turbine)					
Natural gas	760	36	negligible	650	0.19
Coal (w/ scrubber)	1,600	34	90	1,080	0.25
Combined cycle (gas and steam)					
Natural gas	520	47	negligible	54	0.10
Coal (w/ gasifier)	1,700	42	99	90	0.20
Pressurized fluidized bed combustion					
Coal, combined cycle	1,200	42	90	216	0.19
Steam-injected gas turbine					
Gas	410	40	negligible	54	0.12
Coal (w/ gasifier)	1,300	36	99	90	0.24
Intercooled steam injected gas turbine					
Gas	400	47	negligible	36	0.10
Coal (w/ gasifer)	1,030	42	99	72	0.20
Advanced fuel cells					
Gas	600-800	50-55	negligible	18-72	0.09-0.10
Coal (w/ gasifier)	1,000-1,500	45-52	99	36-126	0.17-0.19

Author's note: all values except the conventional steam plant are optimistic, but achievable, numbers in the literature; those for conventional system reflect average values for existing plants. Efficiencies are calculated on a higher heating value basis, the US convention.
Source: Fulkerson, Judkins, and Sanghvi, 1990, p. 87.

For a more detailed comparison of the costs and environmental performance of alternative coal and natural gas-fired power plants, see Pearse 1991, Appendix 4.1.

Methane-rich gases derived by anaerobic decomposition from landfills, sewage treatment plants, and biomass wastes can be used to generate electricity in gas power plants. Bangalore City uses sewage-derived gas to operate a retired airplane engine for power production (Pai, 1989). The 10 tons of slaughterhouse waste generated in Calcutta produce a methane-rich gas (Jash and Ghosh, 1990). A pilot digester is under construction and, if used for generating electricity could produce about 300 kW on a continuous basis, though peaking usage may be preferable. An EPRI (1986) study shows that a 500-kW landfill gas-based power plant with spark-ignited piston engines would have a capital cost of about

$250/kW (1986). Since the raw material is a pollutant, not only is the fuel input free but the environmental costs are negative as well.

Fuel Cells

One electric generation technology not considered by Reddy or Pearse, but noted in Table 6.4, is the fuel cell. A fuel cell converts the chemical energy of a mixture of a hydrogen-rich fuel and an oxidant (typically air) directly into electricity. The efficiency of a fuel cell is not limited by the second law of thermodynamics since the transformation is between two high quality energy forms without passing through heat. Conversion efficiencies thus can be high. Fuel cells currently being developed are likely to compete with other technologies, as Table 6.4 suggests. Fuel cells are modular in construction, and demonstration plants range from a few kW to several MW. This feature and the high efficiency at part load makes the fuel cell suited to decentralized applications. It is completely silent and generates very little pollution, and as some of the environmental costs become internalized, it will find increasing use. For an excellent review of fuel cells see Blomen (1989).

It may be possible to use fuel cells with renewable fuels. One phosphoric acid fuel cell operated on biogas being developed at TERI should produce electricity at Rs 2.40/kWh (TERI, 1990b). While this is a costly and unproven route, it offers the possibility of producing up to three times the amount of electricity per unit of dung as the dual-fuel biogas/diesel engine route.

Producer gas generated from the thermochemical conversion of biomass is a mixture of carbon monoxide and hydrogen with other noncombustible gases. The carbon monoxide can be reformed by steam to carbon dioxide and more hydrogen. Thus, a hydrogen-rich gas could be generated from a gasifier and used to operate a fuel cell.

7

Least-Cost Electricity Planning

In the previous chapter we presented a methodology for comparing the costs of generating or saving electricity using the same economic assumptions. Unit costs of power and energy were developed for a number of generation and conservation alternatives. Once the alternatives are ordered in increasing cost, a least-cost electricity supply curve can be generated as follows.

For each of the conservation alternatives, a savings potential is calculated, e.g., as the annual energy savings per end-use device times the population of devices. The supply curve has the cost of electricity on the vertical axis and the cumulative electricity production or savings on the horizontal axis. Thus the vertical axis is the same as in the previous chapter ($/kWh), but now the horizontal axis is quantified as kWh/yr of demand. The process of generating a least-cost supply curve is shown in Figure 7.1. The starting point of the curve is on the horizontal axis at the demand in the base year. The curve starts as a vertical line of height equal to the cheapest supply or conservation option (technology T1); it then proceeds horizontally on an amount equal to the generation or savings potential of that technology (in TWh/yr). This horizontal extension beyond the base demand corresponds to what part of future demand growth can be met by technology T1. The curve then climbs vertically to an ordinate that corresponds to the next cheapest technology (T2). Again it proceeds horizontally to the extent of the savings or generation potential of T2. This process continues, proceeding along a staircase of increasing cost until it crosses the energy goal (TWh/yr). In Figure 7.1 the T3 step crosses the energy goal. Thus the goal is met most economically by pursuing technologies T1 and T2 to their maximum potential, and T3 until the goal is met. The energy goal is a demand projection for a future year. This future demand must be calculated as a "frozen efficiency" scenario, i.e., the efficiency of end-use devices must be assumed to remain constant at their values in the base year. Thus, only those conservation measures that are found to be cost-effective will enter the least-cost mix of conservation and supply investments.

Figure 7.1
Cost-Supply Staircase (Least-Cost Planning)

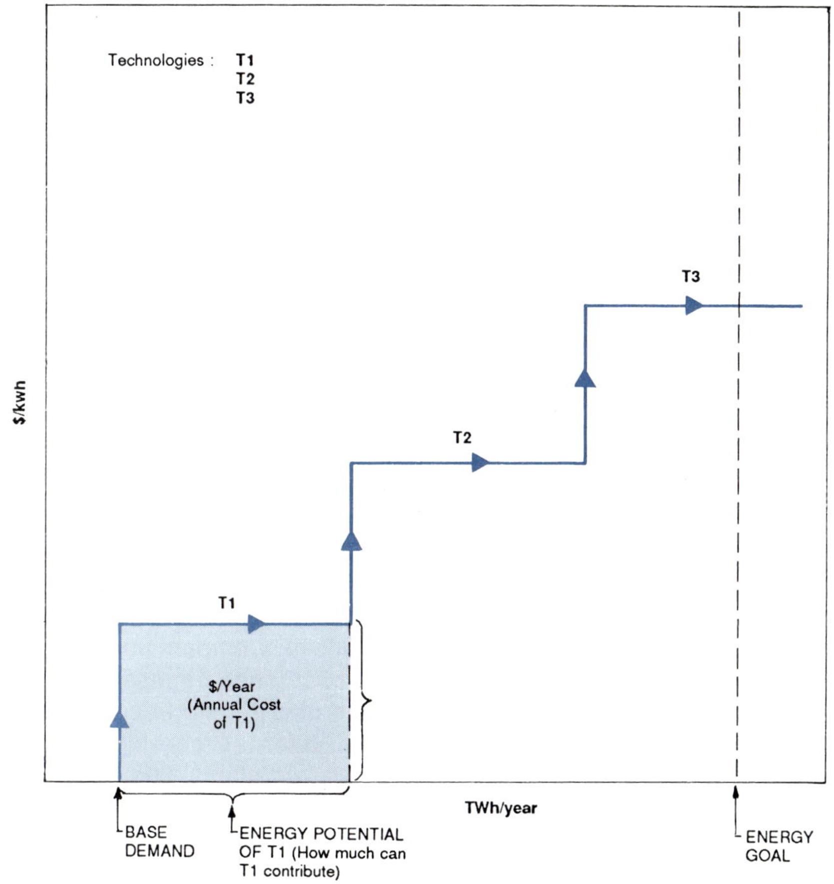

Source : Reddy

The Karnataka Study

The method outlined above was evolved by Reddy, who has used it to develop a least-cost electricity plan for Karnataka State. The future yearly demand can be established using demographic data and macroeconomic projections. In keeping with a development focus, however, Reddy determined the year 2000 demand for electricity with the following additional constraints:

- installation of electric lights in all households in Karnataka,
- use of electric irrigation up to the limit imposed by the groundwater potential,
- establishment of decentralized energy centers in villages, and
- promotion of industries to increase employment.

None of these development objectives would result from a simple continuation of present trends in demand growth. Reddy calls his approach a Development-Focused End Use-oriented, service-directed scenario (DEFENDUS). He projects a year 2000 frozen efficiency demand for Karnataka of 26.6 TWh compared to a 1986 (base year) demand of only 13.5 TWh. Using the comparative costs of electricity conservation and supply alternatives presented in the previous chapter, the least-cost supply curve is shown in Figure 7.1. (The unit cost figures are not exactly the same in this figure as in the data presented in the previous chapter, presumably due to different methods used by Reddy and this report.)

Efficient motors, improved pump sets, and small hydroelectric plants are the least-expensive options, and cost even less than the Karnataka average price of electricity ($0.04/kWh). Compact fluorescent lamps and advanced cogeneration from sugarcane bagasse are only slightly more expensive. Then there is a sharp jump in cost of electricity supply, with four similarly-priced alternatives—biogas, producer gas, natural gas, and large hydroelectric plants illustrated in Figure 7.2. Before a large hydroelectric plant is needed, however, we have already passed the year 2000 frozen-efficiency demand. Coal-fired power plants and solar water heaters are more expensive, while nuclear plants are the most expensive option. Note that Reddy's estimate of nuclear plant electricity cost is in the middle of the range of estimates corresponding to different assumptions. The actual costs could be even higher.

One curious result of this least-cost plan is that the three conventional power supply alternatives—nuclear, coal, and large hydroelectric—are not needed at all to meet year 2000 needs. Figure 7.3 compares the least-cost supply curve with one which is more

Figure 7.2
Karnataka Cost-Supply Staircase

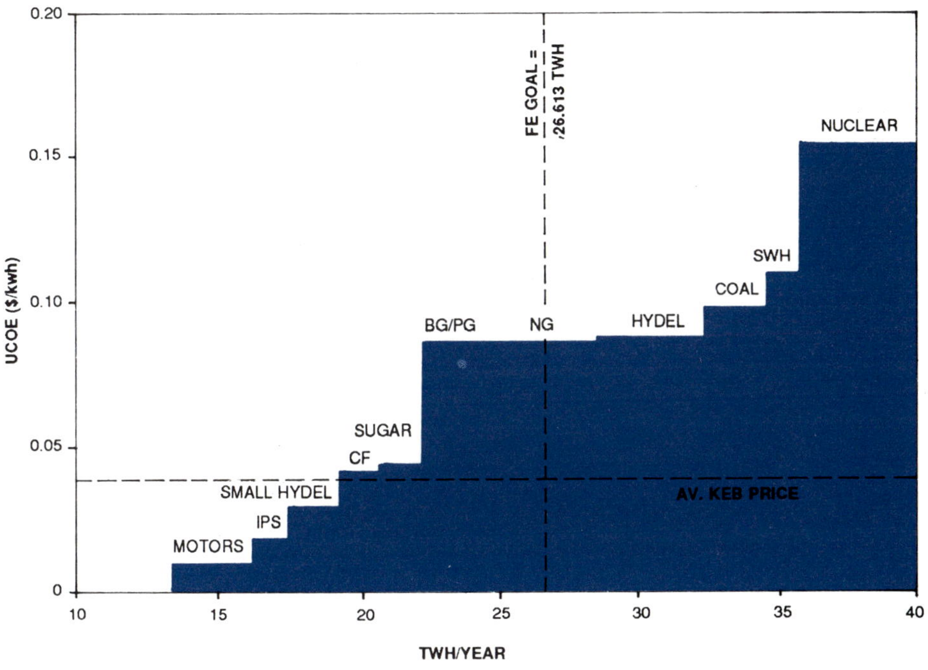

Source : Reddy

Figure 7.3

Planning Approaches (Maximum-Cost vs Least-Cost Planning)

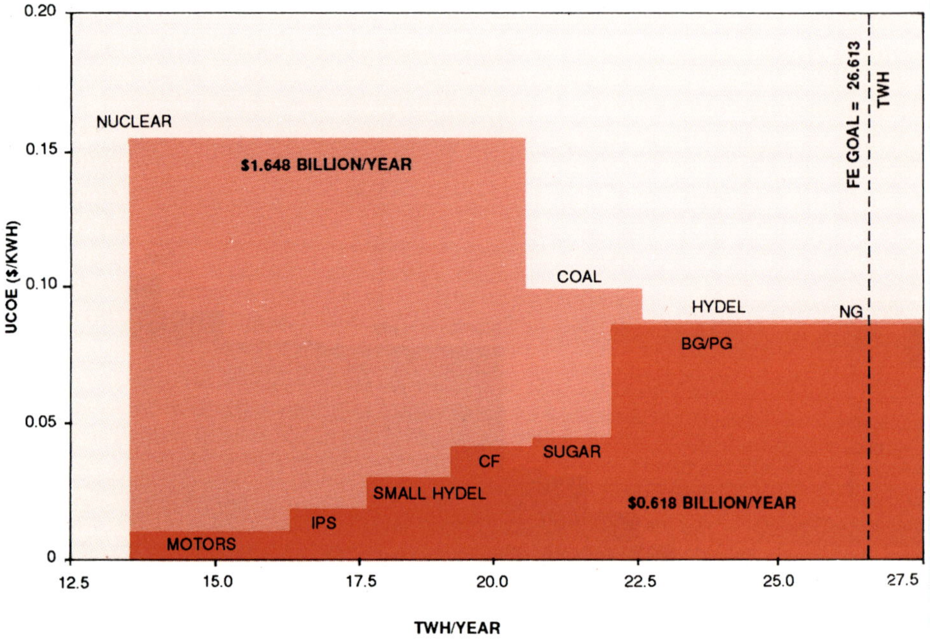

Source : Reddy

representative of standard planning and favors nuclear power, followed by coal and large hydroelectric plants. Working towards the same frozen-efficiency scenario based on a development focus, Reddy found that the least-cost option requires investment of $0.62 billion a year, only 38 per cent of $1.65 billion a year required by the conventional approach.

The frozen-efficiency demand can be met at least cost with a 51 per cent contribution from efficiency improvement, 43 per cent by decentralized generation, and only 7 per cent from central-station power plants (natural gas). If we subtract the contribution of efficiency improvement from the total frozen-efficiency demand of 26.1 TWh, we find that the real demand for electricity that needs to be supplied is only 17.9 TWh. A Karnataka State plan projected year 2000 demand at 47.5 TWh (Reddy and Goldemberg, 1990, quoting a government report).

On the impact of the development focus Reddy and Goldemberg observe that:

> *Fifty-nine percent of the reduction in the energy requirement is achieved through a better-directed development focus: a society that improves the lot of the poor prudently needs less energy than one that makes no such dent in poverty.*

Implications of the Karnataka Study

Reddy's analysis identified two electricity conservation measures to be very cost-effective—efficient motors and improved pump sets. Our analysis of end uses for all India also identified these as the most cost-effective. We considered several other measures—energy-efficient incandescent lamps; and energy efficiency in fluorescent lamps, street lighting, fans, and air-conditioning. Of these we found energy-efficient fans, refrigerators, and air conditioners (all at time of initial purchase in preference to a standard-efficiency unit) were also very cost-effective (cost of saved energy $0.02-0.04 or less). Two areas require urgent development work in order to obtain energy savings without adverse side effects—electronic ballasts and fan speed controllers, both of which introduce harmonics into the power line and lower the (shape) power factor.

The savings potential, estimated by the measures we have considered, is summarized in Table 7.1.

The costs and savings, and thus the cost-effectiveness, of individual conservation measures are not likely to vary much among countries. Thus, the comparative economics of alternative conservation and supply options are likely to be similar. However, they will

Table 7.1
Cost Effective Electricity-Saving Opportunities[a]

End Use	Consumption TWh in 1989-90	Savings Potential[b] TWh/yr, in 5 to 10 yrs
Lighting	29	16.5
Residential (excluding lighting)	12	23.5[b]
Agricultural pumping	32	11
Public water and sewage pumping	3	?
Industrial motors	75	12
Transport (trains, trams)	4	?
Commercial (excluding lighting)	7	?
Industrial (excluding lighting and motors)	2	?
Total use	**183**	**63**

a For further disaggregation, see estimates in individual chapters.

b Savings potential in 10 years, time exceeds current consumption assuming continued rapid growth in the demand for household appliances; this figure also includes savings from ventilating fan improvement in all sectors.

differ in the potential from a given measure, since this will depend on the relative magnitudes of the different end uses. Thus, a detailed analysis of energy end uses is essential to target particular areas for conservation programs.

In our analysis, many conservation measures and alternative generation technologies have not been considered, or their cost-effectiveness has not been quantified. These include street lighting, water supply and sewage pumping, electrolytic processes (conservation), and photovoltaics and fuel cells (supply). Some of these options will no doubt enter the least-cost electricity scenario, further reducing the costs of providing electricity services.

Two fuel-switching options—rural lighting with electricity instead of kerosene lamps, and pumping water with a fuel instead of with electricity—were shown to be economically advantageous from a social perspective, although they do not directly fit into the supply-curve for electricity. These and other alternatives, such as industrial cogeneration, will need to be compared to other options on the basis of life cycle costs.

A method for comparing all options, including fuel switching and cogeneration, to develop an economically optimal energy plan has been devised by Bhatia and recently applied by Mongia et al. (1990). This method uses linear programming to minimize costs given certain constraints. While this leads to a globally optimal solution, the transparency of the supply-curve approach is lost. We expect that both methods will play their part in defining a country's energy strategy.

8

Transport, Energy and the Environment

Transportation systems account for a significant amount of petroleum use and pollutant emissions. Unless appropriate steps are taken, population increase in developing countries is likely to worsen both problems. Solutions include measures to reduce the demand for transport through better urban and transport planning, improved communications, and more energy-efficient and less polluting vehicles. Based on successful experiences in industrialized and developing countries, several policy measures were suggested to develop and implement appropriate transport strategies in developing countries.

The energy used in motorized transport is virtually all liquid fuels derived from petroleum. Most of the petroleum used worldwide is consumed in the 500 million road vehicles, a number that is growing faster than the population (Bleviss and Walzer, 1990). Since 1970 there has been a 4.7 per cent annual increase in the number of cars and a 5.1 per cent increase in the number of trucks and buses.

The dependence of transport on petroleum fuels has a significant impact on the world oil demand and supply. For oil-exporting developing countries, the growing domestic demand for motor fuels reduces their availability for export, and reduces export earnings. This is likely to be a problem for those countries with large populations, e.g., Indonesia, Mexico, Egypt, Nigeria, and others. The problem is much more serious for oil-importing developing countries. They are faced with the dilemma that more and better transport is necessary to support economic growth and development, yet, more and better transport burdens the economy with more petroleum imports (Geltner, 1985). There are also substantial differences between industrialized and developing countries in the distribution of petroleum fuels used in transportation as shown in Table 8.1.

Motorized transport also has a significant negative impact on the environment. Local air pollution results from the emission of lead, carbon monoxide, nitrogen oxides, sulfur dioxide, and particulates. Transport is the main source of airborne lead pollution. Nitrogen oxides and hydrocarbon vapors react in the atmosphere in the

Table 8.1
Distribution (Per cent) **of Petroleum Fuel Usage in Transportation OECD Countries, India and Thailand (in per cent)**

	OECD	India		Thailand[a]	
	1979	1979	1988	1979	1986
Gasoline	60	12	18	39	25
Diesel and Fuel Oil	30	79	76	42	52
Jet Fuel	10	9	6	19	16

a The balance of transportation fuel in Thailand is LPG which increased from 1 per cent to 6 per cent in 1986.

a Sources: 1. Organization for Economic Co-operation and Development (OECD) data – summarized from Note 2.5.3.3-A (Appendix II) of Goldemberg et al., 1988.
2. India 1979 data from Fig. 3.16 of Goldemberg et al., 1988. (see Annex 2).
3. India 1988 data from Reddy, 1990 (see Annex 2).
4. Thailand 1979 and 1986 data from Phiu-Nual, 1989, Table 1

presence of sunlight to produce ozone. A number of these gaseous pollutants exacerbate respiratory and other health problems (see Table 8.2). Confronted by this problem in the 1960s, the US launched a program calling for drastic reductions in the emissions of pollutants. However, because of the increase in the number of vehicles and other factors, this program has only succeeded in keeping pollutant emissions constant (see Figure 8.1). The exception is lead, as unleaded gasoline has gradually replaced the leaded variety in spark-ignited automobile engines.

Nitrogen and sulfur oxides generated by the combustion of fossil fuels in transport and in power plants lead to acid deposition that damages lakes, farmlands, and forests over a wide region, sometimes crossing national boundaries. Finally, carbon dioxide emissions from the combustion of fuels has contributed to an increased atmospheric concentration of this gas that is likely to change the global climate through the greenhouse effect. Although the severity or timing of the impact is uncertain, the potential consequences have become a worldwide concern. Of the carbon dioxide produced by burning fossil fuels, around 14 per cent comes from vehicles. Figure 8.2 shows the amount of carbon dioxide per unit of fuel energy for a variety of fuels. Another impact on the global climate derives from the refrigerants used in vehicle air conditioners. The chlorofluorocarbons persist in the atmosphere a long time and deplete the stratospheric ozone layer, leading to increased ultraviolet radiation on the earth's surface.

Table 8.2
Selected Atmospheric Pollutants

Pollutant	Principal Human Sources	Effects	Remarks
Carbon dioxide	Fuel combustion for heating, transport, energy production.	No direct effect on people. Over time, may lead to increase in earth's temperature.	Normal constituent of atmosphere. Essential to plant life.
Carbon monoxide	Incomplete fuel combustion (as in motor vehicles).	Deprives tissues of oxygen. People with cardio-respiratory diseases more sensitive.	Contribution of natural sources small. Tobacco smoking is more damaging to humans than exposure to traffic.
Sulphur dioxide	Burning of sulphur-containing fuels like coal and oil.	Combined with smoke, increases risk and effects of respiratory diseases. Causes suffocation, irritation of throat and eyes. Combines with atmospheric water vapor to produce acid rain. Reduces crop yields. Leads to acidification of lakes and soils. Corrodes buildings.	
Suspended particulate matter	Smoke from domestic, industrial and vehicular sources.	Possible toxic effects depend on specific composition. Aggravates effects of sulphur dioxide. Reduces sunlight and visibility. Increases corrosion.	Chemically, a most diverse group of substances. Natural sources include dust storms, volcanic eruptions and sea spray.
Oxides of nitrogen	Fuel combustion in motor vehicles and furnaces. Forest fires.	Possible increase in acute respiratory infections and bronchitis morbidity in children. Produces brown haze in city air. Causes corrosion.	Nitrogen oxide and nitrogen dioxide are the two components.
Volatile hydrocarbons	Partial combustion of carbonaceous fuels, industrial processes, disposal of solid wastes.	React with other pollutants to produce eye irritants (acrolein, aldehydes). Ethylene is harmful to plants. Aerosol particles reduce visibility. May produce unpleasant odors.	
Oxidants and ozone	Emissions from motor-vehicles. Photo-chemical reactions of nitrogen oxides and reactive hydro-carbons.	Cause eye irritation and impaired pulmonary function in diseased persons. Corrode materials and reduce visibility. Ozone is one of the most damaging pollutants for plants.	Mainly derivative products of atmospheric reactions between other pollutants. Ozone is a natural and essential constituent of the upper atmosphere.

Sources: India's Environment - 1982
Preparatory Document 48/8, Conference on the Human Environment, Stockholm, 1972

Figure 8.1

Trends in U.S. Air Pollution Emissions from All Sources Normalized to the 1976 Emission Levels

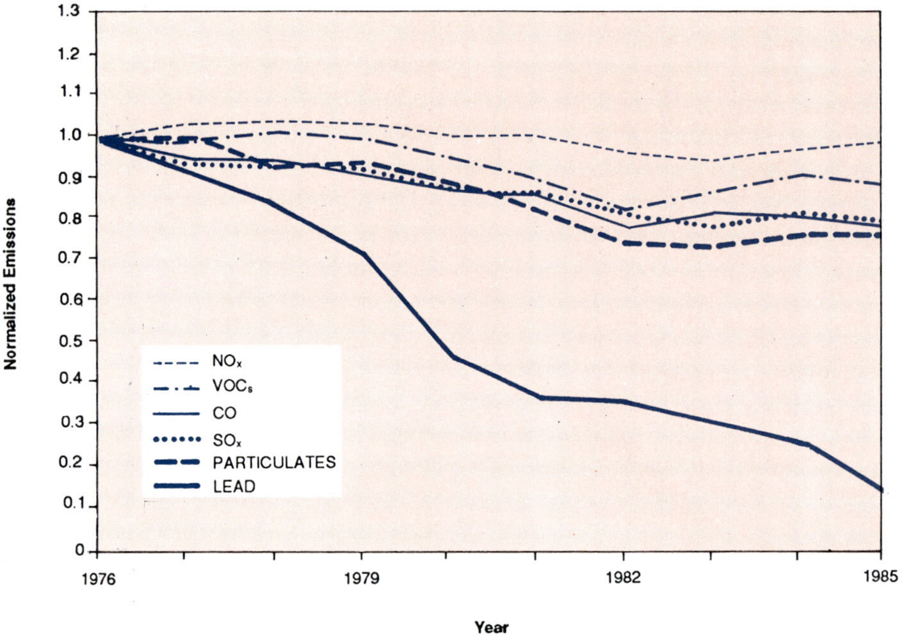

Source : Michael P. Walsh, "Pollution on Wheels," Report to the American Lung Association, February 11, 1988.

Figure 8.2

CO_2 Emissions from the Processing and Use of Various Fuels (In Kilograms of CO_2 Released Per Gigajoule (GJ) of Fuel Energy Used)

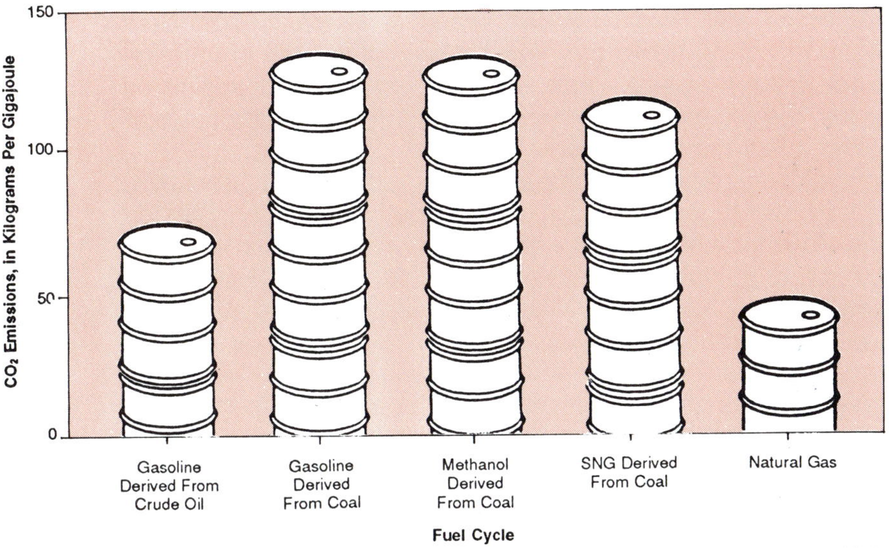

Here it is assumed that crude oil is refined to give gasoline and diesel fuel with 10% refinery losses; gasoline (plus liquid propane gas and synthetic natural gas as byproducts) is derived from coal via the SRC-II process; methanol is derived from coal via the Texaco gasifier process; synthetic natural gas is derived from coal via the Lurgi dry ash process.

The combustion of fossil fuels also leads to water pollution. The relationship between the amount of fuel used in transportation and the risk of accidental release of petroleum in the oceans is highlighted by Greenpeace USA's announcement following the Exxon Valdez' accident in 1989: "It wasn't his (the ship captain's) driving that caused the Alaskan oil spill, it was yours" (cited in Boyle, 1990). Less spectacular or concentrated but larger in magnitude are the routine leaks from tankers in high-traffic shipping lanes. Used engine oil dumped into the ground ends up in the water table and contributes to the pollution of streams and rivers.

Several environmental problems are specific to transport. Accidents in transit are a major cause of death and injury. Related problems are vehicle noise, congestion, and implications for land use. The increased number of vehicles has led to congestion and increased travel times. Industrialized countries that have attempted to deal with these problems through the construction of expressways have had to dedicate much urban land to roads and parking areas. This reduces the aesthetic quality of urban areas, causing people to live far away, increasing transport requirements—a vicious circle in which congestion catches up with increased vehicle capacities of roads.

In less-developed countries, increased road capacity has generally not been able to keep up with substantially higher population growth rates, and emission controls have been minimal so that both congestion and air quality have steadily deteriorated. Transport policy in developing countries (consisting of projecting per cent growth rates on the numbers of cars, trucks, buses, etc.) is neither environmentally sustainable nor should be passed off as development. A comprehensive transport policy should consider energy and environmental aspects as well as the demand for transport.

Need versus Demand

Transport is not one of the basic human needs, but it may be required to satisfy the other needs—food, shelter, clothing, etc. We should not limit ourselves to studying the pattern of end uses that constitutes the present demand for transportation and to looking for alternative technologies for satisfying these end uses. We should also consider the factors that contribute to the present demand. The actual demand for transport depends on urban design, productive arrangements, technology, etc.

If people live far from their work, some form of mechanical transport is needed to get back and forth. If products are consumed far from where they are manufactured, they need to be transported. The distribution and marketing of food and other necessities affect the distances that must be travelled and the form of transport needed. The housing density affects the distances that must be travelled for social and other purposes, and determines the viability of certain mass transit systems. The choice of transportation modes available has significant energy and environmental implications.

The factors affecting transportation demand are not independent of each other nor static in time. The industrial revolution created not only wage labor but also the need for travel to and from work. Urbanization has led to people living in larger urban units where the distances travelled for work, shopping, and services are usually much larger. The growth in popularity of the automobile has led to urban sprawl in some countries, increasing distances travelled, fuel requirements, and emissions.

There have been positive changes as well. The development of the postal system permits letters and documents to carry information, eliminating the need for many personal visits. Yet, a hundred and fifty years after the first postage stamp was issued, the postal systems of many developing (and even some industrialized) countries leave a lot to be desired in speed and reliability. The telegraph, telephone, and more recent developments in communications technology have the potential for reducing the need for personal delivery of information. Technological advances in the industrialized countries have permitted telephone services of higher quality, reducing costs at the same time. Long-distance telephone costs have fallen considerably while advances in the related field of computation have permitted additional communication services (modem, facsimile) to be added to the traditional voice transmission of the telephone network. While these services are available in developing countries, their cost remains high. A long-distance telephone call from Calcutta to Bangalore cost Rs 42 ($2.32) per minute in December 1990, while a second class round-trip rail ticket only about Rs 350 ($19.36). The quality of service, too, leaves a lot to be desired. In Calcutta, a grave and a poem in the office district commemorate the Dead Telephone, and the annual day of remembrance of deceased ancestors also had a telephone receiver being bathed in the Holy Ganges.

While communication can substitute for transportation in many tasks, information can also improve the efficiency of the transport system. Satellites and meteor bursts (Stix, 1990) can help track freight movement in trucks, reducing waiting times and empty returns. Computerized information systems for forward, ongoing, and return reservations facilitate long distance travel. These have been in use for air travel for many years and are also in use for railway reservations in India.

One problem with public transit systems, especially in developing countries, is the lack of available information about routes. For instance, Mexico City has an excellent network of eight subway and two light rail transit lines, with maps readily available. Urban street maps show these lines, facilitating their use. Appropriately, many of the stations are nodes for transfer to and from buses, minibuses, and vans. However, there are no guides for these systems, neither on a citywide basis nor at the interchange stations. While word-of-mouth information is satisfactory in villages and small towns, it fails to be reliable in an urban area of almost 20 million people. Travel to unknown parts of the city by public transport is often hampered by lack of information. An exception in public transit information is provided by the Greater Buenos Aires urban area, described in Appendix II.

The demand for passenger and freight transport can be characterized through origin-destination surveys, for both urban and interurban travel. These surveys lead to the quantification of demand by travel modes, from which the energy consumption (and pollutant emissions) can be estimated. Such characterization is an important starting point for any projection of future demand and alternative strategies for reducing the need for transport (means) without sacrificing the ends currently met.

Table 8.3 shows aggregations of fuel use by type and major transport category, while Table 8.4 shows fuel consumption by vehicle type, both for Thailand. Transport planning requires more detailed information, as illustrated in Table 8.5 for passenger transport in the Netherlands. This study reveals, for instance, that 44 per cent of the total number of trips and 51.7 per cent of the passenger-km are for social and recreational reasons, and over a third of the total passenger-km are for trip distances greater than 50 km. Such characteristics need to be taken into account in making changes to transportation systems.

Performance Indices

The energy efficiency and other aspects of transport systems may be characterized by a variety of performance indices which are discussed below.

Energy. Since transport is not an end in itself, one index of its efficiency is the energy consumed in transport relative to the level of economic activity in a country. Economic activity is generally measured by the gross domestic product (GDP) and expressed in US dollars. Since the cost of goods and services not traded internationally is not the same in dollars in different countries, the GDP is now frequently adjusted according to purchasing power. Table 8.6 shows the transport energy intensity in tons of oil equivalent per thousand dollars of purchasing power parity GDP for selected countries. The data are not up-to-date, but of the countries shown, India has the lowest transport energy intensity. The data in Table 8.6 do not, however, include animal or human energy used in transport, a significant part of the total in India and other low-income countries. The transport energy intensity should include animal-powered transport in terms of the fodder input to the animal.

While the transport energy intensity indicates how much transport energy is being used in relation to a country's economic level, it does not reveal the physical efficiency of vehicles. Energy used can be decreased by reducing the demand for transport as well as by making vehicles more efficient. Physical efficiency may be estimated for the entire transport system or some part of it, but the energy used per unit of transportation, (e.g., MJ/passenger-km or MJ/ton-km) may be used to characterize the entire passenger or freight transport in a country, or that of a particular mode such as freight in long-distance trucks. The physical efficiency of vehicles is expressed as km/liter (used in many developing countries), liters per 100 km (in the European Economic Community) and miles per US gallon (in the US). In this report we will use both km/liter and miles per US gallon (mpg).

Table 8.3
Breakdown of Transportation Fuel by Type Thailand, 1986, in Millions of Tons of oil Equivalent (toe)

Mode	Gasoline	Diesel	LPG	Jet Fuel
Road	1.69	3.42	0.45	–
Rail	–	0.08	–	–
Inland Waterway	–	0.04	–	–
Domestic Coastal	–	0.03	–	–
Aviation	–	–	–	1.12
Total	**1.69**	**3.57**	**0.45**	**1.12**

Source: Phiu-Nual, 1989, Table 4.

Table 8.4
Fuel Consumption of Road Transport by Vehicle Class
Thailand, 1986

Vehicle Class	Fuel Consumption (million toe)[a]
Passenger vehicles	
Motorcycles	0.27
Motor tricycles	0.01
Cars and taxis	1.30
Light commercial vehicles	0.10
Light buses	0.35
Heavy buses	0.78
Subtotal	**2.81**
Goods vehicles	
Light trucks	0.89
Heavy trucks	
6-wheeled	0.73
over 10 wheels	1.11
Subtotal	**2.73**
Total	**5.54**

a toe - tons of oil equivalent
Source: Phiu-Nual, 1989, Table 5

Table 8.5
Survey of Transport Performance in the Netherlands in 1984

Number of journeys, total = 13.996 billion

Journey purpose Work/school	(% of total) shopping/services	commercial	social/recreational
23.7	28.9	3.4	44.0

Number of Passenger-km travelled, total = 137.0 billion pass. km

Modal split bike/walk	(% of total) public transport	motor vehicle	other
11.8	12.2	75.1	0.9

Distance class 0-5 km	(% of total) 5-20 km	20-50 km	>50 km
13.2	27.2	24.4	35.2

Journey purpose Work/school	(% of total) shopping/services	commercial	social/recreational
29.8	12.8	5.6	51.7

Source: Netherlands Agency for Energy and the Environment cited in Boyle, 1990.

Table 8.6
Transport Energy Intensity in Selected Countries[a]

Country	Energy Consumption[a]
USA	0.37
Canada	0.31
FR Germany	0.13
France	0.13
Japan	0.10
UK	0.16
Spain	0.16
Greece	0.14
Venezuela	0.27
Argentina	0.19
Portugal	0.14
Brazil	0.12
Mexico	0.17
Turkey	0.14
Thailand	0.13
Kenya (1970)	0.11
India (1976)	0.08
Indonesia	0.11

a 1977 transport sector (passenger and freight) energy consumption in tons of oil equivalent per thousand dollars of purchasing power parity GDP (1970 dollars). Arranged in descending order of per capita GDP.
Source: Geltner, 1985.

Economics. Transport economics can be characterized in a manner similar to energy. Globally, it can be the proportion of the GDP dedicated to transport, including the manufacture and operation of roads, tracks, airports, docks, and vehicles. The costs of passenger and freight transport can be characterized as $/passenger-km or $/ton-km. Finally, individual vehicles or categories of vehicles can be characterized by $/km or $/seat-km. The latter indices (such as $/passenger-km or $/km) usually consider only the vehicle operating costs and do not include the infrastructure costs which may be shared among various transportation modes and vehicles. This often makes it difficult to apportion the true costs of a specific transportation mode or to compare among alternatives.

Environment. One important environmental burden arising from the use of transport is the emission of gaseous pollutants. The overall impact of a country's transport system can be characterized by the total emissions of the various pollutants, or this figure can be normalized by the purchasing power-adjusted GDP. The environmental impact of a given transport mode may be characterized by the weight of emissions of the different pollutants per passenger-km or ton-km. Finally, the emissions characteristics of individual vehicles or vehicle types may be expressed as emissions per km, e.g., grams per km of carbon monoxide.

Among the problems of characterizing emissions from vehicles is the difficulty of measurement. As there are several pollutants emitted by each vehicle, rating the emissions of different vehicle types or transport modes may require giving relative weights to the different pollutants based on their health, ecological, or climatic impact. There are no simple and reliable measurement procedures and weighing factors for different pollutants. These should be developed given the important connections between transport and (at least) urban air quality. The ultimate index of the success of an air pollution control program is the concentration of the pollutants. This is routinely measured in many countries including in the Third World.

Speed and equity. Two other features of transport systems that need to be quantified are speed and equity.

The speed of passenger transport may be characterized by the average time taken for a specified "typical" journey. One of the car's attractions is the reduced travel time relative to public transit, especially in urban and suburban conditions. Nevertheless, despite an extensive highway network, congestion has so decreased average speeds in Southern California that California is on the brink "of a transportation crisis which will affect the economic prosperity of the State" (Kenner, in Boyle, 1990). As congestion has worsened, average speeds in London have fallen to 10 mph, less than the speeds reached in 1936 and only 3 mph faster than in 1912 (The Guardian, 1989 in Boyle, 1990).

Long before the "transportation crisis" provoked by congestion, Illich (1974) noted that if we add the time one needs to work to cover the expenses of the automobile and the time required to service the automobile to the time the car is actually driven, "the typical American male devotes more than 1,600 hours a year to his car." For the 7,500 miles of travel generated by the car, the "average speed" works out to less than five miles per hour. As the Red Queen of Alice in Wonderland might have said, we have been driving faster and faster to go nowhere at all.

Motivated by frightening environmental trends, one Dutch group developed a "trend-breaking" scenario for the future of passenger transport in the Netherlands. By increasing the railway network (already one of the best in the world) by 30 per cent and public transport capacity by 25 per cent, the scenario targets total passenger km travelled in cars to decrease from 75 per cent to 27 per cent, while the public-transport capacity increases from 12 per cent to almost 50 per cent over 20 years (cited in Boyle, 1990) period. One important aspect of the trend-break scenario is that the time advantage of the private car largely disappears (see Table 8.7). The trend break numbers are based on the assumption that measures such as introducing the 90 km/hr speed limit, closing residential areas for through traffic and city centers and natural areas to all traffic, are all implemented; and that the capacity of the trunk road network is reduced.

Table 8.7 is shown in some detail to illustrate a variety of indices by which transport is characterized. Indices such as these may be used to compare alternative passenger transit systems in developing countries as well.

Table 8.7
A Number of Features of the Public Transport System in the Netherlands Trend Break Scenario

	1984	Trend Public Transport (car)	Trend Break Public Transport (car)
Costs, DFL/Passenger km	0.34	0.35	0.28
Ave. Occupancy level, %	30	30	38
Ave. Journey 19 km			
time, min	54	54(22)	34(22)
speed, km/h	21	21(52)	34(37)
Ave. Journey 76 km			
time, min.	136	136(61)	88(76)
speed, km/h	34	34(75)	52(60)
Number of Junctions in intercity network	19	19	36

Source: NAEE, 1989, as cited in Boyle, 1990

Another important aspect of transportation systems is equity of access. Illich (1974) observed that the speed of one transport mode can generally be increased only at the expense of another. Increasing automobile speeds require expressways, which are not typically open to slow-moving traffic. More importantly, they impede other forms of transport. For instance crossing an expressway by foot or on bicycle generally requires a considerable detour which reduces the speed of these modes. In developing countries only a prosperous minority own cars and are potential users of expressways. Constructing expressways and other transport systems thus may raise important social equity issues that need to be considered.

The Technology of Vehicles and Fuels

Automobile fuel efficiency. Of today's 500 million road vehicles, around 400 million are cars and light trucks. Although only a small fraction of these are in developing countries, this fraction has been growing and is projected to increase to about 20 per cent of the total by the year 2000 (see Table 8.8). Improving automobile fuel efficiency will be important in developing countries, as it has been in the industrialized ones.

Fuel efficiency improvement potential. Engines lose energy because of fundamental limitations of the thermodynamic cycle, incomplete combustion, friction, piston and valve leakage, etc. The diesel cycle, with higher cylinder temperatures, has a higher thermodynamic efficiency than the gasoline cycle. Long popular for trucks and buses, diesel engines are also being used in automobiles in the US and other countries. For the US, the car population data are cars on the road, as estimated in Energy and Environmental Analysis, Inc., "Highway Fuel Consumption Model," Ninth Quarterly Report to the Office of Policy, Planning, and Analysis, Office of Policy Development, US Department of Energy, Washington DC, February 1983, and earlier reports in this series available in smaller vehicles. A significant number of small pickup trucks sold in Thailand are this type. Stratified-charge gasoline engines, which offer reduced emissions of hydrocarbons and carbon monoxide without a catalytic convertor, also show high fuel efficiencies and have been available in some cars since the mid-1970s. Improved combustion control leading to higher efficiency, higher power, and lower emissions can be achieved by replacing the carburetor with electronic fuel injection.

Fuel-efficient urban vehicles are needed even in congested conditions. To save fuel, Indian drivers often turn off the engine at intersections until the traffic control permits them to continue. This taxes the battery and the starter and does not produce savings in stop-and-go traffic or for short waiting periods. The two major ways of reducing fuel use in these conditions are:

- a device that shuts off the engine when the vehicle is idling, and
- a hybrid electric/fuel propulsion system.

Table 8.8
The Automobile Population – Historical Data and Projections
(Historical Data, millions)

	Developing Countries							Fraction
	Africa	**L. America**	**Asia**	**Subtotal**	**OECD**	**E. Europe**	**World**	**in DCs**
1984	7.34	25.60	11.57	44.59	277.83	22.00	344.50	0.129
1983	7.45	24.78	10.11	42.34	270.82	21.20	334.55	0.127
1982	6.98	23.35	9.92	40.25	258.44	20.08	318.77	0.126
1981	6.64	21.17	9.18	36.99	257.59	18.96	313.54	0.118
1980	6.33	18.70	8.25	33.28	253.01	17.09	303.38	0.110
1979	5.85	17.21	8.01	31.07	247.55	15.31	293.93	0.106
1978	5.26	16.29	6.65	28.20	240.43	14.27	282.90	0.100
1977	4.92	15.50	6.86	27.28	232.71	12.58	272.57	0.100
1975	4.34	13.45	5.52	23.31	216.68	10.10	250.09	0.093
Growth Rate (1977-1984) (%/Year)								
	5.9	7.4	7.8	7.3	2.6	8.3	3.4	
Projections to the Year 2000 (millions)								
MIT Study[a]				123.6	362.8	49.6	536	0.230
OECD Study[b]	16	57	27	100	383	46	529	0.189
Average				111.8	372.9	47.8	532.5	0.210
Growth Rate (1984-2000) (%/Year)								
MIT Study[a]				6.6	1.7	5.2	2.8	
OECD Study[b]				5.2	2.0	4.7	2.7	
Average				5.9	1.9	5.0	2.8	

a MIT's International Automobile Program, "The Future of the Automobile," 1984. For non-OECD regions, the projections in this study are actually for less-developed countries (excluding China) and for centrally planned economies (including China).

b OECD, "Long Term Outlook for the World Automobile Industry," Paris, 1983.

Source: Williams, 1986. For countries other than the US, the car population data are for automobile registrations from Motor Vehicle Manufacturers Association of the United States, "Motor Vehicle Facts and Figures, '85" and earlier years. For the US, the car population data are cars on the road, as estimated in Energy and Environmental Analysis, Inc., "Highway Fuel Consumption Model," Ninth Quarterly Report to the Office of Policy, Planning, and Analysis, Office of Policy Development, US Department of Energy, Washington DC, February 1983, and earlier reports in this series.

Volkswagen developed two prototype vehicles using both concepts (Bleviss and Walzer, 1990).

The Eco-Polo is a test car propelled by a small, two-cylinder diesel engine with an advanced fuel injection system. A device called a glider automatic shuts down the engine when the vehicle is decelerating and restarts it automatically when the driver steps on the accelerator. In European urban and highway driving, the test vehicle's fuel economy is about twice that of most European cars. In Indian conditions, the fuel savings would be proportionately larger because conventional cars are less efficient, the urban proportion of driving would be larger, and the urban speeds would be lower.

Volkswagen's other prototype also uses a diesel engine, a small electric motor, an advanced battery and a clutch that connects the motor to the engine. For urban driving the electric motor usually powers the car. When more power is needed, as signalled by pressing down on the accelerator, the electric motor starts the diesel engine which continues to power the car.

Table 8.9 shows basic performance characteristics of fuel-efficient automobiles in production and a few prototypes.

Economics of fuel efficiency. Some of the alternatives for improving fuel efficiency outlined above may not increase the vehicle cost, such as substituting plastics for steel, and would take place even in the absence of any urgency to save fuel. Clearly other changes will require additional investment. Specialty steels, direct-injection diesels, continuously-variable transmissions, hybrid engines, etc. are likely to increase the initial cost of the vehicle.

In 1983, von Hippel and Levi estimated the costs of adding various available fuel-saving technologies to a gasoline Volkswagen Rabbit (Golf). In the initial US model, this 30-mpg (12.7 km/liter or 7.9 lhk) car costs $7,000. The initial cost of the car increases as energy efficiency measures are added, reaching 90 mpg (38 km/lit or 2.6 lhk) at an added cost of $1,725. The extra cost attributed to efficiency improvement may be less than that assumed by von Hippel and Levi. Some of the measures offer other consumer benefits beyond energy efficiency, e.g., more plastics increase product life and decrease repair costs and insurance rates. Turning the engine off during idling could extend engine life, and a continuously variable transmission eliminates the jerkiness in automatic transmission shifting. Moreover, costs of individual improvements are not additive and the developers of the Volvo LCP 2000 expected that in mass production this energy-efficient prototype "would cost the same as today's average subcompact" (Williams, 1986).

The total cost of owning and operating a car includes many items, and is roughly the same over the enormous range of fuel efficiency (see Fig. 8.3). There is no sharp minimum in the curve for total cost corresponding to an "economically-optimal" fuel efficiency. Many years after von Hippel and Levi's work, the notion of a fixed minimum, ingrained through microeconomics textbooks, is hard to overcome and we see statements such as "... even at current low US

Table 8.9
Fuel Economy (1,2) **For Passenger Automobiles**

Car Model Production	Fuel	Fuel (mpg)	Economy (km/1)	Max. Power (kW)	No. of Passengers
In Production					
Daihatsu Charade	diesel	55	23	27	4
	gasoline	52	22	na	4
Fiat Uno ES	gasoline	44	19	na	4
Ford Escort	diesel	45	19	39	5
GM Chevette	diesel	44	19	38	4-5
Honda CRX	gasoline	52	22	45	2
Honda City	gasoline	53	22	na	4
Mitsubishi Minica	gasoline	57	24	na	2-3
Mitsubishi Mirage	gasoline	49	21	na	4-5
Nissan Sentra	diesel	50	21	41	4-5
Nissan March	gasoline	55	23	43	4
Peugeot 205	gasoline	41	17	37	5
Subaru Justy	gasoline	55	23	40	4
Suzuki Fronte	gasoline	59	25	na	2-3
Suzuki Sprint	gasoline	55	23	36	4
Toyota Starlet	gasoline	55	23	57	4
Prototype					
GM TPC	gasoline	61	26	28	2
Brit. Leyl. ECV-3	gasoline	41	17	54	4-5
VW Auto 2000	diesel	63	27	40	4-5
VW E80	diesel	74	31	38	4
Volvo LCP 2000	diesel	63	27	39	2-4
Renault EVE+	diesel	63	27	37	4-5
Renault VESTA2	gasoline	78	33	20	2-4
Peugeot VERA+	diesel	55	23	37	4-5
Peugeot ECO 2000	gasoline	70	30	21	4
Ford	diesel	57	24	30	4-5
Toyota AXV	diesel	89	38	42	4-5

Notes:
- Fuel economy shown corresponds to measurement in an urban driving cycle, as determined by US Environmental Protection Agency procedures. Highway ratings are considerably higher in some areas.
- All production vehicles met the current US standards for air pollutant emissions.

Source: Bleviss, 1988.

fuel prices, all past and projected improvements up to 33 mpg are cost-effective to customers" (McNutt, 1989).

Figure 8.3 shows fuel and total costs for fuel prices ranging from a half to four times the 1981 US fuel prices, and in all cases the total cost is independent of fuel efficiency. This has significant implications towards developing effective policies to promote energy-efficient automobiles. Even relatively high fuel taxes are unlikely to be a

Figure 8.3

The Cost of Driving (In United States Cents per Kilometer and Mile) versus Automotive Fuel Economy

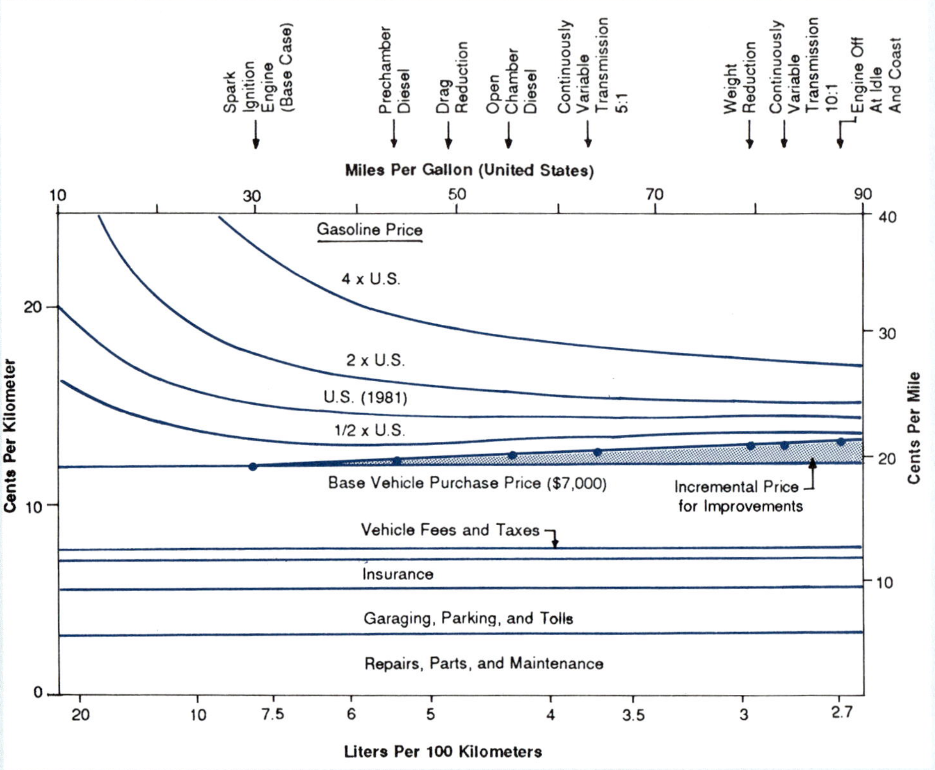

The figure shows that the reduced operating costs associated with fuel economy improvements are roughly offset by the increased capital costs of these improvements, over a wide range of fuel economies.
The indicated energy performance is based on computer simulations of an automobile having various fuel economy improvements added in the sequence shown at the topof the graph. The base case car is a 1981 Volkswagen Rabbit (gasoline version).

Source : F. von Hippel and B. G. Levi, "Automotive Fuel Efficiency: The Opportunity and Weakness of Existing Market Incentives, *Resources and Conservation*, 1983, vol. 10: 103-124.

significant motivation to promote energy efficiency. Minimum efficiency standards will be more effective.

Energy supply implications of pursuing high miles per gallon. Figure 8.3 demonstrates that the user is economically indifferent over a range of auto fuel efficiencies from 30 to 90 mpg. The perspective of the country as a whole is very different: the total amount of fuel required varies by a factor of three as does the foreign exchange requirements for importing oil or oil export revenues foregone. Since the flat minimum applies to all countries, the collective demand for auto fuel will have a significant impact on the world demand and price of petroleum.

Assuming new car fuel efficiencies of 60 to 65 mpg (25.4 to 27.5 km/liter) for the mid-1990s, Williams (1986) estimates an average of 48 mpg (20.3 km/liter) for cars in the year 2000. He notes:

> *If globally there were 533 million autos in 2000 and this average fuel economy, global fuel use by the automobile would be some 6.7 million barrels of oil equivalent per day, 2.9 million barrels per day less than in 1982.*

Emissions. With few exceptions, measures to improve the fuel efficiency of cars also reduce the emissions of pollutants. Reduced friction, transmission losses, rolling resistance, and aerodynamic drag all reduce the amount of fuel burned to provide traction. Pollutant emissions, including carbon dioxide and lead, are reduced roughly in the same proportion.

A number of pollutants are generated in the combustion process and engine modifications affect the rate these pollutants are generated. Some of the ways of reducing energy losses in the engine involve improvements in the combustion process. More complete combustion, such as in a stratified charge engine, reduces carbon monoxide and hydrocarbon emissions. But if the combustion temperatures are increased, more nitrogen oxides are generated. Catalytic convertors in the exhaust stream can remove nitrogen oxides, but their use restricts engine design. Switching fuel and engine type also affects emission characteristics of cars. Diesel engines generate more nitrogen oxides (because of higher combustion temperatures) and significantly more particulates. Thus, in some instances a tradeoff between emissions and efficiency in engine design and fuel choice may be involved.

Improving fuel efficiency and reducing emissions. Most cars sold in developing countries are designed in industrialized countries; some are made there. India's popular Maruti car is essentially a Japanese car assembled in India with many locally made components. Bleviss and Walzer (1990) suggested that any effort to improve the energy efficiency of cars and small trucks should be concentrated in the industrialized countries. However, not all technologies may be suitable for developing countries. McNutt notes:

> *In countries without an infrastructure to maintain sophisticated fuel economy technology (e.g., electronic controls, fuel injection, catalysts, advanced drive trains), a regulatory program that promotes technology-based fuel economy improvement may not be a good idea. Smaller (and lower cost) vehicles with high fuel economy due to their size may be more appropriate than high technology (but higher cost) vehicles.*

We have also noted above that reduction of aerodynamic drag may be much less important while engine-off/on, idle, or coast and other features to reduce fuel consumption and pollutant emissions may be more relevant.

Efficiency of other vehicles. Since cars and small trucks dominate road vehicles, and account for the lion's share of fuel use in transportation in industrialized countries, much greater attention has been paid to improving their fuel efficiency. The situation in developing countries may be very different. Table 8.10 summarizes a few statistics relevant to transport and communications for India from 1981 to 1988. Trucks, buses, and two- and three-wheeled vehicles are a significant part of the total vehicles, and their fuel efficiency must be addressed.

Trucks and buses. Trucks and buses add up to more than half the number of automobiles (cars, jeeps, and taxis) in India and, because of their larger size and longer distances travelled per vehicle, they consume much more fuel. Over three quarters of all the petroleum used in transportation in India is in the form of diesel (Table 8.1) used principally in trucks and buses. In 1978, trucks and buses accounted for 37.5 per cent of petroleum used for all purposes in India (Goldemberg et al., 1988, Table 3.31, citing Reddy, 1985).

Opportunities for saving energy include switching from gasoline to diesel engines and increasing the efficiency of diesel vehicles through improvements in the engine, lubrication, transmission, rolling resistance, and aerodynamic drag. For long-distance trucks, the reduction of aerodynamic drag is important. For urban buses, with much stopping and going, devices that recuperate braking energy will reduce fuel use and pollution emissions. Since trucks and buses use

Table 8.10
Transport and Communications in India, Selected Data

	1981	1985	1987	1988	1988/81
Population (million)	690	766	781	797	1.16
Index No.					
Industrial workers	441	661	719	775	1.76
Fuel, power	416	614	632	669	1.61
Net domestic product (billion Rs)					
Total	1,101	2,078	2,308	2,596	2.36
Transport, commercial	180	385	435	494	2.74
Road (thousands)					
2 & 3 wheelers	2,528	5,121	7,658	9,083	3.59
Cars, jeeps, taxis	1,117	1,546	1,895	2,079	1.86
Buses	154	211	247	269	1.75
Good vehicles	527	782	967	1,054	2.00
Others[a]	847	1,346	1,580	1,999	2.36
Accidental deaths	28.4	39	43.8	48.4	1.70
Rail					
Passenger km, billion	209	241	256	269	1.29
Freight, billion ton-km	158	206	223	231	1.46
Goods wagons, 1000s	401	360	354	347	.87
Route length, 1000 km	60	61	62	63	1.05
Coastal Shipping					
Vessels, 1000s	65	95	127	138	2.12
Gross tons, 1000s	300	296	438	443	1.48
Air (domestic)					
Passenger km, billion	446	746	811	870	1.95
Freight, billion ton-km	6	9	10	12	2.00
Communications					
Telephones, 1000s	2,785	4,044	4,408	4,741	1.70
Trunk calls, million	169	218	229	230	1.36
Post articles, billion	9.84	12.23	12.75	13.54	1.38
Post and telecom Revenues, billion Rs	9.1	15	23.3	30.7	3.37

a The "others" category for road vehicles includes farm and special vehicles.

Table 8.11
Vehicle Emission Inventory for Delhi, 1983-1984

	Emissions, 1,000 tons/yr		Vehicle Type
	CO	HC	NOx
2 and 3 wheelers	29.3	18.3	N.A.
Cars and Jeeps	25.4	5.3	1.7
Taxis	2.8	0.5	0.2
Diesel Vehicles	2.8	0.5	19.3

a major part of the fuel in many developing countries, more research and development effort needs to be directed towards them, particularly in the larger developing countries.

Two- and three-wheeled vehicles. In India, two- and three-wheeled vehicles are much more numerous than cars (Table 8.10), and despite much lower fuel consumption per kilometer, account for a significant proportion of total gasoline use. India is perhaps a major exception, where much motorized private passenger transport is on two-wheeled vehicles and many taxis are three-wheeled vehicles. These operate on two-stroke engines that are both inefficient and highly polluting. Table 8.11 shows some estimates of total emissions of carbon monoxide, hydrocarbons, and nitrogen oxides from vehicle types in Delhi. Thus, a significant effort to reduce emissions and increase efficiency is warranted in these vehicles.

One Australian company has been developing improved two-stroke engines for use in cars (Bleviss and Walzer, 1990). Inherent features of two-stroke engines promote incomplete combustion, making improvement difficult. Since this is a problem almost unique to India, significant research and development in these vehicle categories should be conducted there, and an attempt made to resolve the problems, even if it is by using four-stroke engines.

Railways. Railways are much more efficient than road vehicles. Thus, they are less expensive, particularly for long-distance freight and passenger transport. Nevertheless, the fraction of freight carried by railways has decreased over the years in India and in many other countries, for a variety of reasons including quality of service and fuel-pricing policies. There has been a 13 per cent decrease in the number of "goods wagons" in Indian railways between 1981 and 1988 (Table 8.10). Given the enormous energy penalty associated with rail transport, steps should be taken so that railways may recapture their share of the market. Appendixes III and IV discuss this problem in the context of a holistic solution of India's rural electrification, cooking fuel, and oil problems.

Technological changes can help to improve the efficiency of railway vehicles. The most significant step to improve fuel efficiency in Indian railways has been replacing coal-fired steam locomotives by diesel and electric ones. Table 8.12 shows the energy intensities for steam, diesel, and electric traction for long-distance passenger and freight transport. One consequence of the shift in locomotives is that although rail freight increased from 87.7 to 152.5 billion ton-km and rail passenger transport increased from 77.6 to 193 billion passenger-km between 1960-61 and 1978-79, total energy consumption by the railways decreased from 8.6 to 6.7 million toe per year during these periods (Dunkerley et al., 1987). By 1986-87, total energy consumption decreased further to 5.6 million toe per year (TERI, 1989a).

Table 8.12
Energy Intensity in Non-Suburban Passenger and Freight Transport (Indian Railways, 1982-1983)

Type of Locomotive	Energy Intensity (toe/million gross t-km)
Steam	60.8
Diesel	3.5
Electric	3.4

Source: Railway Board, cited in Dunkerley et al., 1987.

The shift from steam to diesel or electric is a one-time savings. Some other energy-saving measures are being implemented by Indian Railways. These include reducing wheel/rail friction, reducing rolling resistance of wagons, coaches, and locomotives, (including reducing the coach weight), improving the aerodynamic profile of the locomotives, and improving the conversion efficiency of locomotives (TERI, 1989a). The latter category includes a prototype chopper device which has demonstrated electricity savings of up to 34 per cent in Bombay suburban services. Regenerative braking, where the kinetic energy "loss" during braking is used to generate electricity can save 3-5 per cent of energy used, and will improve the power factor and reduce line losses by about 20 per cent.

Between 10 and 12 per cent of the energy is used for purposes other than traction. There, too, significant savings have been demonstrated, e.g., by replacing mercury vapor by high-pressure sodium lamps for outdoor lighting.

Shipping. As in the case of the railways, shipping is more energy efficient than other modes, and energy savings can be brought about by increasing freight and passenger transport in ships, and by improving the ship's fuel efficiency. In Thailand, in 1986, inland and coastal shipping consumed only about 2 per cent as much diesel as road transport (see Tables 8.13 and 8.14), while carrying 11 per cent of the freight ton-km, and an unspecified part of passenger traffic (Phiu-Nual, 1989). (Railways used a little over 2 per cent as much diesel and carried 10 per cent of the freight of road transport.) In India, the gross tonnage of coastal ships increased by 48 per cent between 1981 and 1988 (Table 8.10). Increases in the number of road vehicles in the same period were considerably higher, suggesting a decrease in the relative amount of freight transport in coastal shipping.

Table 8.13
Breakdown of Transportation Fuel by Type
(Thailand, 1986, millions of tons of oil equivalent, [toe])

Mode	Gasoline	Diesel	LPG	Jet Fuel
Road	1.69	3.42	0.45	–
Rail	–	0.08	–	–
Inland waterway	–	0.04	–	–
Domestic coastal	–	0.03	–	–
Aviation	–	–	–	1.12
Total	1.69	3.57	0.45	1.12

Source: Phiu-Nual, 1989, Table 4.

Clearly, water transport is not possible everywhere, but its use relative to road transport promises large energy savings. Ships can also be made more efficient by shifting from coal to diesel fuel, and by improvements in the combustion process.

Air transport. Perhaps the most spectacular transport fuel savings already realized have been in air transport. The gradual replacements of propeller and turboprop by jet aircraft, and larger and more efficient jets replacing smaller ones have considerably improved fuel efficiency. Between 1971 and 1983, the energy intensity in India's passenger air transport almost halved—from 0.083 to 0.149 liters per available

Table 8.14
Fuel Consumption of Road Transport by Vehicle Class in Thailand, 1986
(million tons of oil equivalent, [mtoe])

Vehicle Class	Million toe
Passenger vehicles	
Motorcycles	0.27
Motor tricycles	0.01
Cars and taxis	1.30
Light commercial vehicles	0.10
Light buses	0.35
Heavy buses	0.78
Subtotal	**2.81**
Goods vehicles	
Light trucks	0.89
Heavy trucks	
6-wheeled	0.73
over 10 wheels	1.11
Subtotal	**2.73**
Total	**5.54**

Source: Phiu-Nual, 1989, Table 5.

seat km (Dunkerley et al., 1987). Although more recent data are not at hand, the energy intensity should have fallen further since then, as aircraft substitution continued.

Alternative fuels. Most current alternatives to petroleum fuels are not economically viable.

Ethanol from fermented cane sugar was cost-effective in Brazil when international petroleum prices were high in the early 1980s. The vast land areas needed to produce sugar cane or other energy crops are unlikely to be available in most developing countries, which must dedicate all available agricultural land to feeding an increasing population.

Use of the alternative petroleum fuel, compressed natural gas, is increasing particularly in countries facing a shortage of liquid fuels. Gasoline engines can be readily modified to operate on natural gas. The cost of the modification and one or two gas cylinders limits cost-effectiveness to high-use vehicles in urban areas, e.g., taxis. Since the availability of natural gas depends on a pipeline and special fuel delivery facilities, gas availability also limits the suitability of this as an alternative fuel. Natural gas can substitute for a part of the fuel in diesel engines. To operate purely on natural gas, the diesel engine must be converted to a spark-ignition (essentially a gasoline) engine (LFMG, 1990; Tulyanon, 1989).

The attraction of alternative fuels is mainly to their reduced emissions of pollutants. In a number of countries (e.g., Kenya and Zimbabwe), ethanol is used as an octane enhancer in place of lead. This also reduces the petroleum fuel requirement without the need for engine changes or dedicating large land areas to fuel alcohol production since the alcohol content of the fuel is at most 20 per cent.

Methanol offers the potential of reduced emissions. Methanol can be made from coal, natural gas, or from biomass feedstocks through a variety of proven technologies (Mills and Ecklund, 1987). To meet air quality standards, Southern California calls for all fleet vehicles to run on methanol by 1993, and effectively bans all gasoline-driven vehicles by 2007.

These alternative fuels also produce carbon dioxide (see Figure 8.2), except when based on a biomass feedstock that is grown renewably. The only transportation fuels that promise virtually no carbon dioxide emissions are hydrogen derived from solar energy (Ogden and Williams, 1989) and electricity derived from sources that do not involve burning carbon (specifically hydroelectric and nuclear power). As the cost of solar cells continues to fall, hydrogen derived from electrolysis is likely to become economically competitive with other transportation fuels provided that some of the environmental costs of using alternative transportation fuels are internalized (e.g., through a tax on the carbon content of fuels).

Any attempt to introduce alternative fuels must be coordinated with vehicle and engine manufacturers, and will require significant changes in the fuel delivery infrastructure.

Transport Strategies and Policies

Sustainable transport strategies may be grouped into five categories:

- reducing the needs for transportation while satisfying the demand,
- shifting to more energy-efficient, environmentally-benign modes,
- improving the efficiency of vehicles and reducing emissions,
- shifting to alternative transportation fuels where feasible, and
- considering holistic solutions that resolve other problems together with those of transportation (for example, see Appendixes II and IV).

Achieving the goals implied by these strategies will require a variety of policy measures:

- Reducing the needs for transportation involves urban design and the development of improved communications.
- Shifting to more efficient modes presupposes the existence of an adequate transportation infrastructure.
- Improving the efficiency of vehicles, reducing emissions, and shifting to alternative fuels will require substantial research and development, especially in areas that are not of primary importance to industrialized countries.
- Implementing measures to reduce fuel use and pollutant emissions will require efficiency labels, standards, and fiscal incentives.

Each of these policy measures is discussed below.

Urban planning. Urban form has a significant impact on transportation energy use. Developing countries with rapidly growing populations may reduce future transportation needs through appropriate urban designs. Two urban form factors are particularly relevant: the degree of concentration and the degree of region-wide integration of functions (van Til, 1980). Higher density increases the feasibility of public transit, increases its usage, and appears to be associated with higher vehicle occupancy levels for both private and public transportation (Destor, Miloff and Haas, 1979).

A diversified-integrated city permits people to live near places of employment. This was typical of the commercial cities of 18th century (van Til, 1980), but the changes in industrialized societies of this century have led to declining self-employment, and increasing employment in large organizations. This has tended to separate urban functions, with people preferring not to live near large factories. The post-industrial transition with the increased importance of information and services in the economy (Benniger, 1986) reverses the trend, permitting more diverse urban forms. Developing countries should take this into account, rather than continuing to separate industrial, commercial, and residential areas. Instead of having dirty industry located far from residential areas, these societies can choose to locate clean convivial industry and living areas closer together. This would reduce work-trip distances and energy use, and by making urban areas more environmentally attractive, it will reduce the need to live in far-flung suburbs or to flee the city at weekends (Boyle, 1990).

Even when residential, commercial, and industrial areas are not well integrated to reduce transport needs, one way of improving a transport system is an urban design with many mini-centers instead of a single downtown area. Travel distances are reduced as people converge on the mini-center instead of a central business district, as in many US cities. The congestion and air pollution from unidirectional rush hours are avoided. The mini-centers of a multi-node city may themselves be connected by appropriate mass transit systems, depending on traffic density. In a multi-node design, the only distinction between a single city and an entire region is one of scale.

Nepal's capital region, Kathmandu Valley, is a good example of a three-node city, of three urban areas separated by about 10 km. Each urban area is of high density, with narrow streets, and two of the three areas have no motorized traffic within their limits. The three areas are connected by buses or trolley buses.

Kathmandu Valley's concentrated urban areas were originally intended to minimize construction on fertile agriculture land, and while this remains a valid design objective, it can serve as a model for tomorrow's cities. The distribution of nodes within a city will probably be determined by geographical features, e.g., cities along a river (Calcutta) or a coast (Bombay) may have nodes linearly spread along the shores. In more inland areas without any obvious directional advantages (e.g., New Delhi and Bangalore), nodes distributed on radial lines and concentric rings around a center may be desirable.

Communication infrastructure. If communication is to substitute for certain kinds of travel, the communication systems must be greatly expanded and improved. Improving telephone and related communications infrastructure will require significant investment in upgraded technology and possibly improved management. Besides the benefits in reduced communication times, significant transportation energy savings may be realized.

Besides the availability of technology, administrative changes are also needed to permit transactions by telephone. In India, for example, a request for a replacement of a liquid petroleum gas (LPG) cylinder requires a personal visit. In industrialized countries and some developing countries (e.g., Mexico) even banking transactions can be made by telephone. Salary is deposited into and regular monthly bills automatically debited from a bank account. In the US, mail and telephone purchases by credit eliminate many trips. Goods are delivered by mail or by a trucking company that optimizes its trip length, reducing vehicle and personnel time, and energy costs. Developing countries should work to remove bureaucratic obstacles that prevent greater use of communications to substitute unnecessary travel.

Transportation infrastructure. While improved urban design and communications can reduce the need for transport, clearly they will not eliminate transport altogether. The energy and environmental impact of the remaining demand for transport will depend on its design. In industrialized countries with little population growth, the urban pattern changes relatively slowly, and transportation system design must assume a fixed urban form. In developing countries, on the other hand, integrated transportation and urban design can potentially provide excellent transportation services while keeping energy consumption and environmental impact modest.

The city of Curitiba, capital of Parana state in southern Brazil, is an example of a radius-and-ring urban design with integrated mass transit (Lerner, 1989). The arterial roads include corridors dedicated for mass transit, lanes for slow-moving vehicles, and expressways in each direction. The mass transit in Curitiba consists of express buses.

Subways and elevated rail, which provide separate corridors, are very expensive. Even the cheapest light rail systems costs $10 to 15 million per km. Dedicated busways, which are also used in Porto Alegre (Brazil) and in Abidjan (Ivory Coast), usually cost no more than $2 to 3 million per km (Philips, 1991, citing Urban Transport Working Group Strategy Paper, UNCTADA II, World Bank, December 1990). Apart from lower initial cost, another advantage of dedicated busways is that as passenger carrying capacity is exhausted, vehicles may be upgraded, e.g., first with articulated buses and later with a light rail system (Lerner, 1989).

In Curitiba, express bus lanes are complemented by feeder and interdistrict lines. One 20-km express bus corridor was introduced in 1974. Fifteen years later there were five arteries with 70 km of exclusive corridors for buses, 494 km of feeder, and 157 km of interdistrict lines carrying over 500,000 passengers a day. A single fare includes transfers between buses at interchange bus stations. The shorter routes connect neighborhoods (generally upper middle class) located near the downtown area and generate a surplus which subsidizes longer routes serving lower middle class and blue-collar neighborhoods. Given that dedicated bus lanes, like railway lines and expressways, impede the movement of pedestrians, bicycles and other slow-moving vehicles, the equity issues of the bus lanes needs to be considered. Curitiba has addressed it at least partially through an internal subsidy to low-income residents. In 1989, 50 tickets a month (considered to be a typical transport demand) cost only about 10 per cent of a minimum wage salary, one of the lowest proportions in Brazil.

Curitiba land use plans have encouraged residential and commercial construction close to mass transit lines. One consequence of good mass transit combined with urban planning is that although Curitiba residents have one of the highest rates of car ownership (0.2 per capita), it has one of the lowest rate of gasoline consumption per vehicle of any Brazilian city, and the lowest among any comparably-sized city. The Curitiba experience demonstrates the results possible through integrated land use and transportation planning centered around urban mass transit.

The experience in Curitiba in transportation planning and Buenos Aires in transport information systems (Appendix II) may be compared with that of Delhi. Delhi's bus network is based on nine nodes surrounding a central node. Buses do not have dedicated lanes. Although there is a ring railway, its potentially high passenger-carrying capacity is not used—railway stations are not important transportation nodes. The transport information system also needs improvement. Figure 8.4 shows a page from a Delhi guide listing bus routes. With only three points defining a bus route, and keyed map, the guide does not help people get around from intermediate points. Improved urban transport systems are inexpensive and should be developed or improved in all developing countries.

Research and demonstration. Any successful transportation development program should include a strong research and development component. Some of the areas where research will be needed in any society are outlined below.

The present demand for transport and its impact on energy consumption and contribution to environmental problems should be studied and quantified. Urban form and other factors that determine the total transportation demand, use of alternative transit modes, etc. should be investigated. Methodologies applied to industrialized countries suggested in Destor, et al. (1979), Bohi (1981), and NAEE (1989) may be a convenient starting point. Using operations research techniques, origin-destination surveys have been conducted to improve urban bus routes in several Indian cities. The results of previous origin-destination surveys and the ability to conduct similar surveys at short notice may permit additional insight to be gained. Short-term inter-mode elasticities can be determined by comparing number of passengers and fuel consumption before and after fuel price changes, and during strikes affecting one mode. There is little experience with measuring elasticities between transportation and communication but, given its relevance, methodologies for such analyses need to be evolved and applied to developing countries.

Research effort also needs to be directed to technology improvement in selected areas. For domestic industries to produce energy-efficient and low-emission cars, trucks, and buses, some domestic research and development is needed to adapt technologies developed in industrialized countries to local manufacturing

Figure 8.4
A Page from a Delhi Guide Showing Bus Routes

182	Railway Station	Rithala	Maurice Nagar
183	Clock Tower	E'Block Jahangir Purl	Bara Bagh
184	Railway Station	BH'BVlock Shalimar Bagh	Clock Tower
185	Kendriya Terminal	Dr. Mukherji Nagar (W.T.)	University
186	Railway Station	Uttri Pritam Pura	Gupta Colony
187	Palika Kendra	Siras Pur	Shankar Road
188	Kendriya Terminal	Qutab Garh	Old Sectt.
189	Kendriya Terminal	Model Town-III Mkt.	Khalsa College
190	Kendriya Terminal	Dhacca Chowk	Old Sectt.
191	Mori Gate Terminal	Singhu Border	Adarsh Nagar
192	I.S.B.T.	Nathu Pura	Jharoda Dairy Xing
193	Mori Gate Terminal	Khan Pur	G.T.K.R. Depot
194	Mori Gate Terminal	Bakauli Temple	Nangil Punam
195	Railway Station	Siras Pur	Model Town-II
196	Kendriya Terminal	Uttri Pritam Pura	Mandir Marg
197	Railway Station	Raza Pur (Prashant Vihar)	Model Town-II

capabilities, climate, fuel availability, driving conditions, etc. Industries in developing countries are not noted for their research capabilities or interests. Taiwan enacted a mechanism for promoting research by requiring all companies to spend a set percentage of their sales income on research and development or forfeit the difference to the government which will then use it to fund other research (Bleviss, 1989).

At the very least, developing countries will need to develop test procedures and build testing facilities. Given that vehicle emissions testing requires measurements of energy performance, integrated test procedures for energy consumption, emissions, noise, and technical performance (power, acceleration, braking, etc.) are suggested (Bleviss, 1989; McNutt, 1989).

Research and development in developing countries may need to concentrate on trucks, buses, railways, and (in India) on two-and three-wheeled vehicles. Indian experience suggests that it is very difficult to make these vehicles more efficient without greatly increasing cost or reducing flexibility. Nevertheless, their importance requires continued and more serious attention. In this report, we have made scarce mention of animal- and human-powered vehicles.

Energy efficiency labels and standards. Vehicle test data may be used to inform consumers through standardized energy efficiency labels on vehicles. Clearly, one cannot promote the purchase of fuel-efficient cars, motorcycles, etc. if consumers cannot determine relative efficiencies. Secondary indices that are sometimes used, such as vehicle size and engine capacity, are often unsatisfactory.

Fuel economy improvements based on technological improvements can change the traditional relationship between vehicle size and fuel economy: a smaller car may not be more efficient (McNutt, 1989). Labels can be based on measurements of fuel consumption over a specific driving cycle.

Labels alone do not suffice, particularly if energy-efficient models are not offered. To improve the penetration of energy-efficient models of cars, many industrialized countries created targets for energy efficiency. Most successful car programs have the following attributes in common (Bleviss, 1989):

- targets are company-specific, rather than overall national goals, so that each company can monitor its progress;
- targets are regularly increased;
- targets apply both to domestic production and imports;
- targets exist for both passenger cars and light trucks; and
- if targets were originally voluntary, an enforcement option replaces voluntary efforts that fail to materialize.

Industrialized countries have varied types of target. The US calls for fleet-wide fuel economy levels by manufacturer (see Tables 8.15 and 8.16). Japan sets requirements by weight category. Some countries specify a percentage improvement in a given period (Bleviss, 1989). Countries that represent relatively small markets of imported vehicles should coordinate activities with neighboring ones to develop region-wide targets in order to gain support from vehicle suppliers who might otherwise resist a segmented market. For instance, the ASEAN group of countries could adopt a single set of targets for auto and light truck fuel efficiency.

Regarding the US experience on setting standards for fuel efficiency for light trucks, McNutt (1989) notes:

> *The fuel economy program improved greatly between 1975 and 1980, a time of oil price shock, poor fuel economy in available cars, and strong public support for increasing vehicle fuel efficiency. From 1981-88, US prices declined, as did interest in fuel economy from the public, vehicle manufacturers and the government. Support for the regulatory program also declined. Fuel economy standards were relaxed as market trends towards smaller vehicles bottomed out. (See Figure 8.5)*

Figure 8.5
U.S. Vehicle Fuel Economy Trends

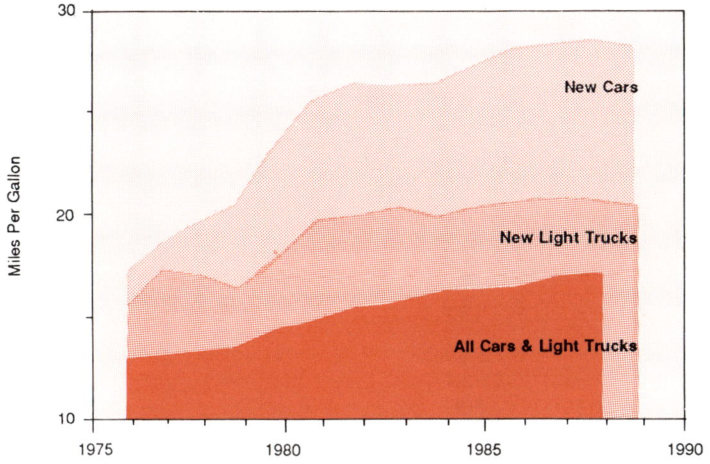

Table 8.15
U.S. Fuel Economy Standards for Cars and Light Trucks, MPG (km/liter)

Year	Cars	Trucks
1978	18.0 (7.6)	–
1979	19.0 (8.1)	17.2 (7.3)
1980	20.0 (8.1)	16.0 (6.8)
1981	22.0 (9.3)	16.7 (7.1)
1982	24.0 (10.0)	17.5 (7.4)
1983	26.0 (11.0)	19.0 (8.1)
1984	27.0 (11.4)	20.0 (8.5)
1985	27.5 (11.7)	19.5 (8.3)
1986	26.0 (11.0)	20.0 (8.5)
1987	26.0 (11.0)	20.5 (8.7)
1988	26.0 (11.0)	20.5 (8.7)
1989	26.5 (11.2)	20.5 (8.7)
1990	27.5 (11.7)	20.0 (8.5)

Source: McNutt, 1989, Table 1.

Table 8.16
Gas Guzzler Tax on New Cars
1986 and Later Model Years (US)

	Fuel Consumption	**Tax**
	> 1.5 mpg	**> 9.1 km/l**
21.5 - 22.5	9.1 - 9.5	500
20.5 - 21.5	8.7 - 9.1	650
19.5 - 20.5	8.3 - 8.7	850
18.5 - 19.5	7.8 - 8.3	1,050
17.5 - 18.5	7.4 - 7.8	1,300
16.5 - 17.5	7.0 - 7.4	1,500
15.5 - 16.5	6.6 - 7.0	1,850
14.5 - 15.5	6.1 - 6.6	2,250
13.5 - 14.5	5.7 - 6.1	2,700
12.5 - 13.5	5.3 - 5.7	3,200
< 2.5	< 5.3	3,850

Source: McNutt, 1989, Table 2

Although corporate-average fuel efficiency standards have worked in the US, they tend to favor manufacturers who specialize in smaller vehicles (McNutt, 1989).

The experience of industrialized countries in automobile fuel efficiency targets needs to be extended to trucks, buses, and two- and three-wheeled vehicles depending on their relevance to the country. To verify that vehicles indeed save fuel as suggested by the label, the test measurements need to be supplemented by data based on actual road performance for fuel economy and emissions.

Targets may need to be developed for components as well. While radial tires have become the norm in industrialized countries, for their improved durability and performance (rather than particularly for fuel economy), their higher initial cost has prevented them from gaining a significant market share in India. The accelerated introduction of radial tires is an essential component of reducing fuel use. Mechanisms for promoting energy-efficient components include fiscal measures (e.g., a differential tax) as well as standards. Fiscal measures for promoting energy efficiency and reduced emissions are discussed below.

Fiscal measures. Clearly, a starting point is to set rational energy prices. Subsidies that lead to more energy-intensive transport modes should be eliminated wherever possible. Because of the relationship of transport fuels to energy demand and pricing for other uses, it may not be easy to eliminate all such subsidies. Appendixes III and IV discuss this problem for India, the energy implications, and a strategy for resolving the problem.

Getting the price right is not enough. Earlier, citing an analysis of the costs of car ownership as a function of fuel efficiency, we noted a "flat minimum" (Figure 8.3). Even at high fuel prices, the total cost of car ownership appears to be virtually independent of fuel efficiency. Higher fuel taxes are not likely to be an effective policy to promote auto fuel efficiency. Higher fuel prices are generally ineffective in promoting fuel savings according to various studies estimating the long-term fuel price elasticities of demand. Boyle (1990) notes:

> *The implications of (estimated price) elasticities would be that an annual fuel tax increase of 5 per cent to 17 per cent would be required to reduce overall car fuel consumption by 20 per cent in 15 years (Jessup, 1989), leading to an eight-fold increase in current North American prices between 1993 and 2005.*

With minimum fuel efficiency target regulations, and a "gas guzzler" tax, the US has improved its vehicle fuel economy almost to European and Japanese levels without resorting to high fuel taxes (Bleviss and Walzer, 1990).

While fuel taxes appear to be an ineffective mechanism for promoting fuel efficiency, vehicle taxes and ownership fees may be useful. Many countries have fees that increase with weight, cylinder capacity, horsepower or a combination (generally with decreasing fuel economy). To promote fuel efficiency and technological innovation, however, the tax should be based directly on fuel efficiency tests (such as those used for labeling new models), instead of the surrogate variables which may not be accurate indicators of fuel consumption. The US "gas guzzler" tax is levied only once (at original purchase) on all vehicles that are below a specified level of fuel efficiency. The lower the fuel economy, the greater the tax (see Table 8.16). Moreover, the threshold level below which tax is assessed also increases over time.

Ownership fees have successfully discouraged the purchase of "gas guzzlers." US auto manufacturers have endeavored to improve fuel economy of their largest models to avoid the tax. In New Zealand, which has imposed a highly progressive sales tax on cars based on cylinder capacity, average engine size dropped from 2.0 to 1.68 liters between 1975 and 1981. The more progressive a fee is, the more significant will be its impact. In New Zealand, the tax more than doubles from top to bottom of the spectrum (Bleviss, 1989).

Subsidizing mass transit (e.g., rail transit and diesel fuel) appears ineffective. Measurements of cross-price elasticities between transport modes (e.g., auto vs. transit or truck vs. rail) are typically very small (Geltner, 1986). Thus, little can be saved by subsidizing an energy-efficient mode. Subsidies are widespread, and, if unavoidable, Geltner suggests that the subsidy, fuel price, and tax policies be complementary. For instance, if mass transit is subsidized, it should be done as far as possible with realistic prices at the margin. For example, the government may issue a "lifeline" or "basic needs" supply of tradeable transit stamps or coupons to urban or low-income residents, but then should require transit operators to pay full diesel costs and raise fares accordingly. This way not all diesel users are subsidized. If diesel fuel must be subsidized for mass transit, then truck tires, tolls, and license fees or a tax on freight revenues could be increased to partially offset the advantage to diesel trucks.

A variety of rationing measures has been used to control transportation demand, motivated by energy, air pollution, or congestion problems. One way is by limiting fuel sales, e.g., no sale on Sundays, or by issuing coupons limiting the amount of fuel that can be bought. Alternative measures tend to control vehicle use, e.g., cars with certain license plate numbers are not allowed to be used on certain days of the week.

While rationing may be necessary in emergency situations, it should not be used as a policy measure for reducing transportation demand. It is administratively difficult and open to abuse. To get round the one-day-a-week ban on driving, Mexico City families owning multiple cars use the "other" car on the days one is banned; others purchase new cars while keeping their old one to use once a week. Even without these responses, drivers shift all nonessential activities to other days of the week with little reduction of automobile use.

Another form of rationing is to limit vehicles from congested areas. This has been more successful in Mexico City where the city center almost always registers lower levels of air pollution than surrounding areas. New York City's Manhattan has also successfully limited automobile traffic through parking restrictions, high parking fees, and other measures that make driving unpleasant. To get around the limited carrying capacity of Calcutta's roads, trucks are only allowed on certain roads late at night.

Any rationing measure should consider the economic costs of restricting traffic. One way of maintaining economic efficiency is by charging fees (e.g., parking fees or tolls in urban expressways) equivalent to the environmental costs of air pollution created by the traffic.

Operations and Management. A variety of operations and management programs can also help to reduce transportation energy use or environmental impact. The Philippines has implemented a number of steps which operations and management programs, summarized below (Gimenez, 1989).

- *Operators, facilities, personnel and equipment.* Educational programs for car and bus drivers and mechanics include a basic energy management seminar, and more than 5,000 people have participated in them since 1978. Establishments consuming more than 2 million liters of fuel oil equivalent a year are required to employ energy managers and to submit energy conservation programs and quarterly consumption reports.

- *Traffic flow improvement.* Heavy and slow-moving trucks are banned during morning and afternoon rush hours from Manila roads. School and working hours were rescheduled to spread out the peak hours. Traffic lights were synchronized, using a system of computers, signal controllers, vehicle detectors, and other elements. Significant reductions in travel time and accident rates have been documented.

9

Bioenergy Alternatives for Cooking

Gautam Dutt and N.H. Ravindranath

Improved stoves have been developed to reduce fuel consumption of and air pollution from traditional biofuels that currently provide cooking energy to half the world's population. Biomass also can be transformed into a liquid, gaseous, or improved solid fuel prior to use in a cookstove. This chapter presents bases for comparing alternative biofuel cooking systems and illustrates the alternatives, with data from cooking trials in a South Indian village

About half of the world's population continues to depend on biofuels—fuelwood, charcoal, crop residues, and dung—for cooking. While households in the industrialized countries and some in the less developed countries (LDCs) have shifted to petroleum fuels and electricity, these options will probably not become available to most rural people in the LDCs. In rural areas, cooking often consumes 70-80 per cent of total energy use and biofuels supply most of the energy. The urban poor also often use biofuels for cooking. In a future post-petroleum era, some petroleum fuels are likely to be replaced by higher-quality gas and liquid fuels derived from biomass.

A cooking system consists of the fuel, the stove, and the cooking vessels. The cook may also be considered as part of this system. We start with the criteria by which alternative cooking systems may be evaluated. We then review alternative bioenergy cooking systems, based on traditional and alternative biofuels. (Cooking with solar energy is qualitatively different, and this renewable option will not be discussed here.) We follow by presenting users' fuel preferences and the energy use and economics of alternative cooking systems, including some conventional fossil-fuel and electricity-based options as reference. Lastly, we consider some environmental consequences of alternative biofuel systems.

Evaluating Cooking Systems

Cooking systems may be compared through many criteria (Masera, 1990). A technical evaluation includes some measure of the system's energy efficiency and economics as well as the stove's suitability for cooking and other needs. An environmental evaluation includes the limits of the fuel resource, health impacts of the cooking system, and environmental consequences of large-scale shifts in cooking system choice. In LDCs, particularly in rural areas, a sociocultural evaluation is essential and must include assessment of the impact on broader development objectives and on cultural integration of alternative technologies. Some of the evaluation criteria are discussed below.

Fuel availability. The choice of cooking systems is often limited by the availability and reliability of fuel supply. The petroleum-derived cooking fuels—kerosene and liquid petroleum gas (LPG)—are often not available at all, or are available irregularly in rural areas and small towns of LDCs. Fuelwood supplies are limited in some areas because there are few trees left, and only lower quality fuelwood (twigs and shrubs) may be available. Kerosene is not marketed at all as a cooking fuel in some countries, LPG in others. Natural gas is rarely available as a household cooking fuel in LDCs. Biogas cooking fuel can only be supplied if sufficient dung or other suitable biomass resources are available. Certain cooking fuels may have infrastructure requirements such as piping, metering, and billing for a gaseous fuel supplied to households.

Stove suitability in pots. The stove of an alternative cooking system must be suitable for the required variety of cooking tasks. In some parts of the world, cooking includes boiling in pots, and baking on flat pans or griddles. Traditional cooking systems perform all cooking tasks, whereas some improved fuelwood stoves are designed for round or flat-bottomed deep pots and cannot be used for grilling or baking. In such cases, even if the improved stove is accepted by a household, some traditional cooking devices will continue to be used.

For instance, typical gas stoves available to Mexican households are not suitable for the preparation of tortillas, the principal cooking task in most rural households. Thus, the few households in a typical village that have access to LPG continue to use fuelwood for tortilla preparation. Urban households, on the other hand, invariably purchase ready-made tortillas and a gas stove is satisfactory.

Other cooking characteristics include speed of cooking, ability to see the fire and control heat output, stove safety, mobility, durability, ease of maintenance and of removing soot from the vehicles (Joshi, 1988). In order to ensure the suitability of an alternative cooking system, a survey of potential users should precede stove introduction, and user surveys should be a part of dissemination programs of alternative systems.

A rural cookstove may serve a variety of noncooking functions, such as heating water or providing space heat. In an open fire, all the heat released is available as space heat. However, houses with open fires have to be sufficiently open to allow air inflow for combustion needs. Most improved stoves reduce available space heat, particularly when used with a chimney. The reduction is desirable in hot countries, but not in colder climates. A chimney in an improved stove removes the smoke from the cooking area. While this has a direct health benefit, the smoke from a traditional open fire may have been used for insect control. With an improved cooking system, space heating, insect control, and other functions may need to be served by other procedures and devices. Noncooking functions should be included in user surveys.

Fuel consumption. Improved energy efficiency, achieved by using an improved stove or an alternative fuel, is usually an important function of an alternative cooking system. Reduced air pollution, another important function, is discussed in the following section. In the process of improving the efficiency of wood-burning cookstoves, standardized procedures have been developed for measuring their fuel efficiency (Volunteers in Technical Assistance, 1982). An improved version was presented by Baldwin (1987) and is recommended here. The procedures can be used or adapted to any cooking fuel. The standards include three types of tests—water-boiling tests (WBTs), controlled cooking tests, and kitchen performance tests.

In the WBT, a measured quantity of water is boiled and simmered while water temperature, time, and fuel use are recorded. WBTs are useful for laboratory evaluations of alternative cooking systems and for developing improved stoves. Test results are expressed as an efficiency. Efficiency may be defined in several ways and, in the commonly accepted procedure, is the heat absorbed by the water in the pot (including the latent heat of water evaporated) divided by the higher heating value or gross calorific value of the fuel and adjusted for the moisture content of the fuel (an important factor in biomass fuels).

Table 9.1 lists the higher heating values of some cooking fuels. Some researchers use lower heating value; many do not specify which one is being used. For a given stove, water-boiling efficiencies depend on pan size, shape, material, and quantity of water. Thus efficiency figures reported by different researchers (or even by the same researcher) are not always comparable. In a given evaluation, all stoves should be compared under identical test conditions.

In the controlled cooking test, a local cook prepares a meal typical of the region where an alternative stove program is being considered. Researchers compare the stoves' specific fuel consumption, which is defined as weight of fuel per weight of cooked food (kg/kg) or energy content of fuel used per weight of cooked food

Table 9.1
Higher Heating Values of Cooking Fuels

Fuel	Higher Heating Values (MJ/kg, except as noted)
Wood, air dried	15.9
Sawdust, air dried	15.9
Wood charcoal	29.0
Biomass briquette (not carbonized)	17.8
Dung cake, air dried	13.1
Biogas	(22.3 MJ/m^3)
Producer gas	(4-6.0 MJ/m^3, at 0^oC and 1 atm.)
Anhydrous ethanol	(24.0 MJ/liter)
Kerosene	43.1
LPG	45.2
Natural gas	50.0

(MJ/kg), the latter being more useful for comparing stoves using different fuels. (Laboratory tests, including water boiling and measurement of high and low power input to the stove, can be used to estimate the fuel consumption in a defined set of water-based cooking tasks [EUT, 1985]. Where a significant amount of the cooking is not water-based, it is harder to estimate cooking fuel use from WBT data.)

Because the controlled cooking test is conducted under controlled conditions, it may not correctly reflect stove fuel consumption in actual conditions. Actual users may not use the stove as the designer intended and thus may not obtain the best fuel economy, stoves constructed as part of a large diffusion program may not conform exactly to critical design dimensions that affect stove efficiency, and the fuel economy realized may reduce fuel-scarcity pressures and encourage increased stove use.

The impact of a cooking system on fuel use is best evaluated by measuring fuel use in households before and after a cooking system change. These measurements, called kitchen performance tests, are done for a statistically meaningful sample of households (often 100 or more), and usually in conjunction with stove user surveys. Since measurements under actual use cannot be too intrusive, they are limited to the quantity (weight or volume) of fuel use per capita per day, converted to energy content, and expressed in MJ/person-day. For natural gas, LPG and electricity, a meter is usually needed to measure cooking fuel use. If LPG is used only for cooking, fuel consumption can be determined from cylinder weight.

Smoke and air pollution. Biomass-burning cookstoves generate a number of air pollutants—particulates, carbon monoxide, nitrogen oxides, sulfur oxides, and a variety of hydrocarbons—that have adverse health effects. Cookstoves based on other fuels generate some or all of these pollutants, and others, (e.g., nitrogen and sulfur oxides) in significant amounts.

Air pollution from cookstoves rarely has been characterized because this requires complex and expensive equipment. One proposed standard method evaluates the stove's performance during a standardized water-boiling procedure in a chamber where the ventilation rate and the concentration of two pollutants—carbon monoxide and total suspended particulates—are monitored (Ahuja et al., 1987; Joshi, Venkataraman and Ahuja, 1987). The fuel used for the water heating task is also measured. These data are used to calculate the:

- emission factor (weight of pollutant generated per unit weight of fuel burnt) for carbon monoxide and total suspended particulates,
- stove efficiency, and
- emissions per standardized task for each pollutant (by normalizing the emission factor by the amount of fuel burnt).

The last indicator considers the effect of changes in fuel use to perform a specific cooking task. This procedure, which is an extension of the water-boiling test for measuring stove efficiency, can characterize stoves using fuels that are not based on biomass.

Economics. Although not all features of alternative cooking systems can be quantified, the alternatives' capital and operating expenses can be compared. Among the economic criteria normally used to evaluate energy efficiency, simple payback and internal rate of return are best for comparing only two alternatives. Cost of saved energy is a popular index that allows comparison of conservation and supply alternatives where one fuel type is being saved or produced. The annualized life-cycle cost (ALC) allows a simple comparison among two or more alternatives with varying stove lifetimes and fuel types. The ALC annualizes the capital investment in the cooking system (I), using a capital recovery factor (CRF, which depends on a discount rate), plus annual fuel cost (FC) and other recurring operations and maintenance costs (OMC). The total is the annualized cost of owning and operating the cooking system:

$$ALC = (I \times CRF) + FC + OMC.$$

The lowest ALC is the least-cost alternative for a given discount rate.

An economic analysis of this type can be conducted using market values of stove and fuel costs where these are commercialized, typically in urban areas. This presents the economics from the users' perspective, provided that the discount rate chosen is appropriate for the user. Market prices of fuel, and sometimes stoves, include taxes and subsidies. Most rural people in LDCs gather fuel and build their own stoves at zero private cost. The economics for the user's perspective can be extended to estimate stove and fuel shadow prices by multiplying local labor rates by the probability of employment and the time taken to build a stove or gather a unit of fuel. If much of the productive activities are outside the monetary economy, these shadow prices may not be meaningful.

An economic analysis can also be conducted from a societal perspective, where internal cash flows (taxes and subsidies) are ignored and external costs are considered. External costs include environmental consequences of using alternative stoves and fuels, and the cost of foreign exchange (e.g., for imported stoves or fuels). Clearly these are hard to quantify and are likely to be region-specific. The societal perspective should include a "social" discount rate. An economic comparison of alternatives from a societal perspective can be an important element in policy formulation.

Social and cultural factors. Technical merits are not the only consideration in how alternative cooking systems are viewed and evaluated by the end user. Saving cooking fuel is often not a priority among the rural population faced with other pressing needs. Changes that alter the kitchen arrangement and thereby the social contact among household members may be rejected. A new fuel or a major change in stove design may be viewed as a sign of progress and a status symbol. The participation of local women in the process of innovation and diffusion of alternative cooking systems is necessary for social and cultural factors to be taken into consideration (Masera, 1990).

Bioenergy Cooking Systems

Three-stone open fire. In many parts of the world, cooking is done on an open fire with the cooking vessel supported on three stones. The three-stone fire was used universally in ancient times, and in many societies it carries a great deal of religious significance.

Three-stone fires are generally used with firewood and crop residues. Through the mid-1970s and the early 1980s, the three-stone fire was reputed to have very low efficiency, around 5 per cent—

based more on speculation than on measurement (Gill, 1987). Later, measurements revealed widely varying efficiencies, sometimes as high as 36 per cent.

Traditional stoves. A variety of enclosed fires or stoves are also used traditionally in many parts of the world. In Asia, these are typically made of mud and are built by the users. Like the three-stone fire, they do not have chimneys and they release the smoke into the kitchen through openings around the pot. They are used with fuelwood, crop residue, or dung. They can be single or multi-pot types. In water-boiling efficiency tests, traditional stoves fare little better than the open fire and efficiencies higher than 20 per cent have not been recorded (Gill, 1987).

Improved fuelwood stoves. Many attempts have been made to improve wood-burning stoves to reduce kitchen smoke levels, energy use, or both. The current wave of interest in improved fuelwood stoves may be traced to the Hyderabad chula (stove) in the early 1950s and the Singer stove developed for Indonesia by the Food and Agriculture Organization in about 1960 (Raju, 1953; Singer, 1961). Both are multi-pot stoves that are built in situ and constitute one of the three main categories of "site-built stoves." The second category is a portable stove, meant to be artisan-built or mass produced. The third category is an insert stove with artisan-made pieces that are installed in houses to form the complete stove.

Site-built stoves can be made from bricks and cement (as in China), concrete (Burkina Faso), mud (India), or a mixture of volcanic ash and clay called lorena (Central America). They typically accommodate two or three pots and have a chimney. In one series of water boiling tests, their efficiency was found to be 15-22 per cent while the three-stone fire was 17 per cent (Baldwin, 1987). The Indian Institute of Science conducted an empirical study of three-pan mud stoves with chimney. Through a sequence of dimensional optimization, the efficiency was raised from 15 per cent to 45 per cent in a stove called the ASTRA ole (Lokras et al., 1983). The precision of certain dimensions is crucial to high efficiency and requires a trained stove builder. Since each stove is site-built, quality control can be difficult. Many site-built stoves, including the ASTRA ole (Fig. 9.1), are sized to fit certain cooking pots. When used with other pots, overall efficiency and the distribution of heat among the pots may be greatly reduced.

Portable stoves. To reduce manufacturing cost and maintain better quality control, some researchers have developed stoves intended to be artisan built. Since these stoves are built in workshops and sold

Figure 9.1

ASTRA OLE - A Three-pot Mud Stove with Chimney (India)

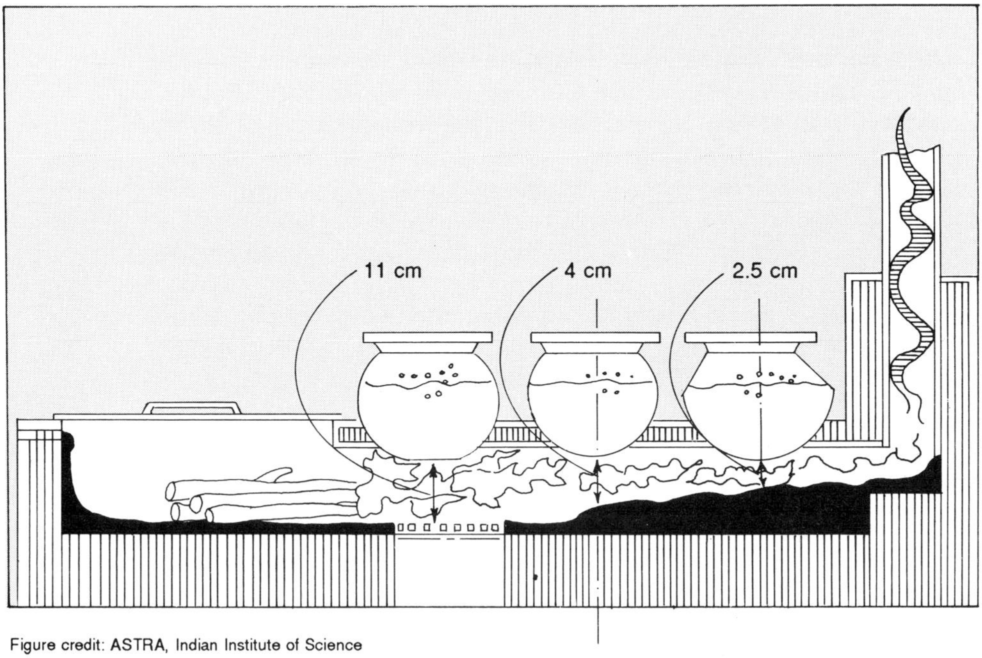

Figure credit: ASTRA, Indian Institute of Science

to the user, they are necessarily portable. Portable stoves are usually made of metal or fired clay. The first major energy-efficient sheet metal stove was designed by Baldwin and coworkers in Burkina Faso in 1983 (Baldwin, 1987). Each stove is dimensionally optimized to suit a single-pot size. Pots larger than a maximum size cannot be used (Fig. 9.2) and smaller pots substantially reduce efficiency. Baldwin recorded efficiencies of 29 per cent in a sheet metal stove of simple design, 36 per cent in a ceramic stove, and 42 per cent using an insulated metal stove. Other metal stoves, including several Indian designs (Priyagni, Tara, and Swosthee), show efficiencies as high as 43 per cent (Baldwin, 1987; Mukunda, Shrinivasa and Dasappa, 1988). These artisan-built stoves have an inherent quality control advantage as critical dimensions are easier to maintain under factory conditions. A sample from a batch of 200 Swosthee 1 kW stoves (model S1) had an efficiency indistinguishable from that of the prototype (Fig. 9.3).

Since metal is unavailable or expensive in some rural areas, while local potters are available, fired clay offers another building material for artisan-built portable stoves. Since the dimensions cannot be maintained as precisely as in a metal stove, efficiencies recorded are somewhat lower, though costs are usually considerably lower. Production tests in Burkina Faso in 1983 showed manufacturing costs of $2.25 for a metal stove and $0.57 for a fired clay stove (Baldwin, 1987).

Insert stoves. An insert stove combines the advantages of artisan-built and site-built stoves. The critical stove parts are artisan-built, usually of ceramics. The user installs these pieces in the kitchen, employing mud. The artisan-built parts help maintain critical dimensions, while site installation permits the construction of a multi-pot or chimneyed stove. Ceramic-insert stoves (Fig. 9.4), both with and without chimney, have been developed and promoted in Nepal (Joseph et al., 1984), Sri Lanka (Amerasekera and Sepalage, 1988), and elsewhere.

Charcoal stoves. Converting wood to charcoal has long been a way to improve fuel quality. Charcoal is a relatively smokeless fuel but generates considerable carbon monoxide. It was favored by affluent households before petroleum cooking fuels such as kerosene and LPG became available. Charcoal is still used extensively, particularly in the urban areas of many African countries.

The efficiency of converting wood to charcoal depends on wood moisture content and the type of kiln used. An extensive study in Thailand tested 13 charcoaling methods including earth mounds, portable metal kilns, and "permanent" kilns (Charcoal production..., 1985). Permanent kilns were found to yield better quality and more

Figure 9.2

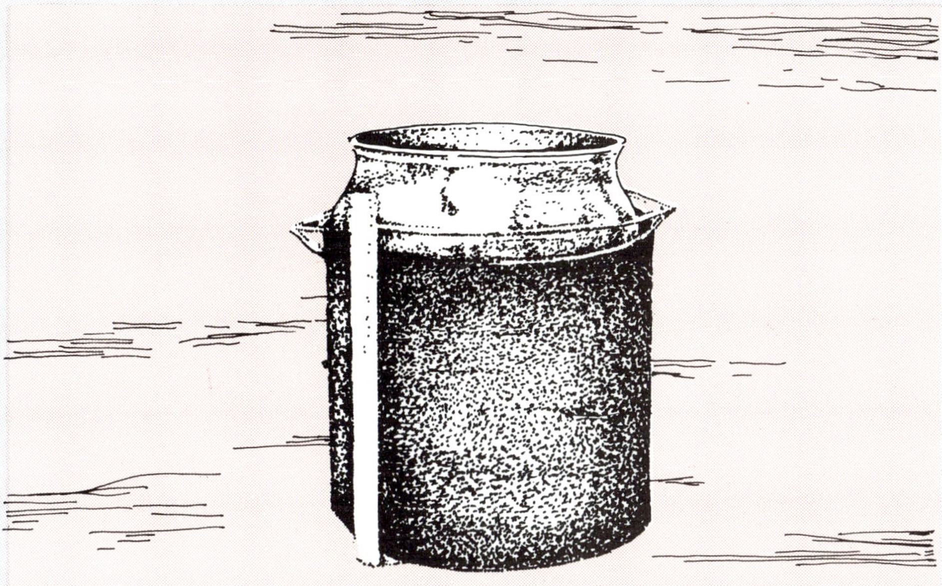

An improved metal stove (West Africa). Note how the pot sits inside the stove – stove and pot must be matched for optimum performance.

Figure 9.3
Efficiency of Two Metal Stoves of 1kW and 2kW Nominal Power

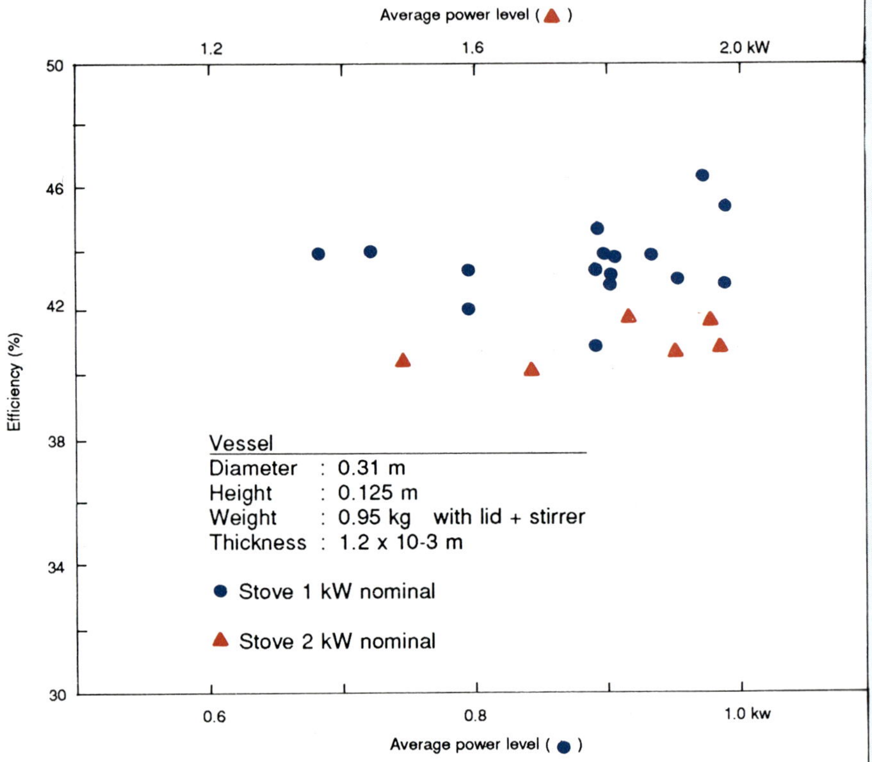

Note: The "2" refers to a second 1kW stove built to the same specifications but in a lot of 200.

Figure 9.4

An Artisan-built Pottery Insert that is User-installed in the Kitchen to Make a Complete Stove

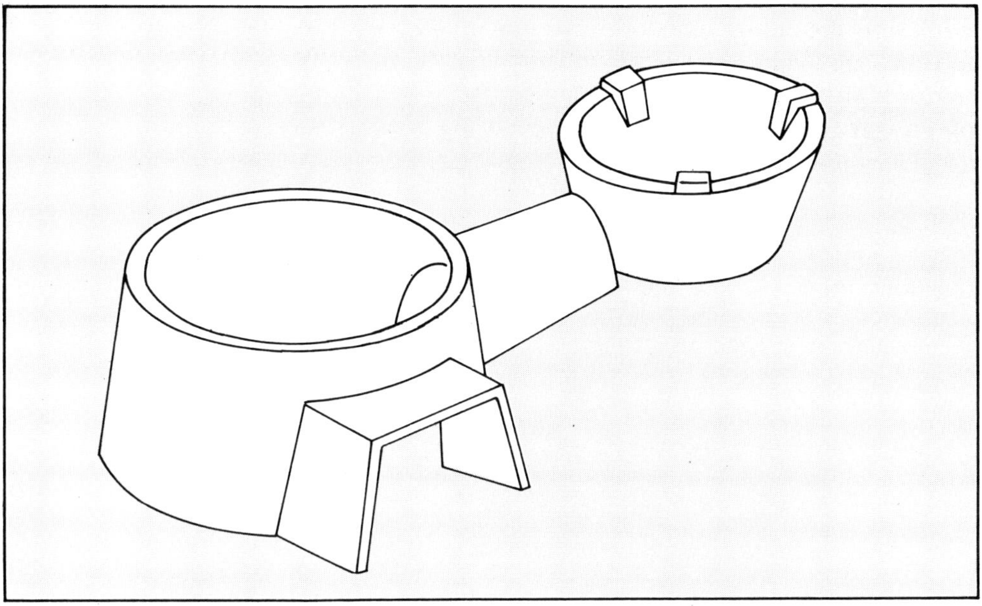

Figure credit : Intermediate Technology Development Group

charcoal than nonpermanent mounds and metal kilns. Selected results are presented in Table 9.2. Careful tending produced much higher yield and efficiency in the traditional earth mound than is normally found (Leach and Gowan, 1987). Earlier we noted a similar result for the three-stone fire. Substantial cost reductions are, however, possible with brick beehive kilns. Charcoal price depends on raw material, transport and labor costs, as well as the conversion technology. Table 9.3 shows charcoal selling prices in selected urban areas in Sudan, Kenya, India, and Haiti.

Figure 9.5 shows a typical traditional charcoal stove. Improved charcoal stoves have been developed in several countries. Kenya's popular improved jiko consists of a ceramic liner and grate and an insulating layer inside a metal shell that reduces heat loss from the stove sides and bottom (Fig. 9.6). UNICEF developed an all-metal, double-walled stove called the Umeme that allows air inflow and power control (Fig. 9.7). Sangen and Visser (1986) tested the comparative performance of 13 charcoal stoves, using a water-boiling test procedure wherein charcoal was periodically charged to keep fuel weight roughly constant. While this is not how people normally load fuel during cooking, the procedure reduces the variability of test results (Bussman, 1988). Important results are summarized in Table 9.4. The power efficiency of most stove models is high, exceeding 40 per cent. Also important is the extent to which the heat output can be reduced to save fuel during simmering. Unlike the fuelwood stove, where heat output can be varied by changing the amount of wood, heat control in charcoal stoves requires varying the flow of air into the combustion area. Four of the 12 stoves tested had no such control, while five of the improved stoves had a low-power burn rate that was a third or more below the high power-burn rate. These stoves (the Thai bucket stove, the Haiti improved stove, the KENGO jiko, the improved coalpot, and the UNICEF jiko) are likely to require the least amount of fuel in actual cooking. At least one of these (the KENGO jiko) has already become a very popular stove.

Stoves for biomass residues. Crop residues and other biomass wastes are alternative cooking fuels. When they have no fertilizer value, their use reduces the problem of waste disposal and the demand for fuelwood. Plant stalks and straw can generally be burnt in traditional fuelwood stoves. We consider the direct use of powdery and densified (briquetted) biomass fuels.

Powdery Biomass. Some forms of biomass—sawdust, rice husk, etc.—are powdery and require a specially-designed cookstove. A common traditional design found in many countries is a "kerosene can," a 16-liter square container that is packed with biomass, leaving

Table 9.2
Conversion Efficiency and Economics of Alternative Charcoal Kilns in Thailand

Kiln Type	Volume m^3	Construction Cost ($)[a]	Yield (Per cent)[b]	Efficiency (Per cent)[c]	Production Cost ($/ton)	Production Cost ($/GJ)
Earth mound	0.7	-	31.1	50.9	145	4.99
Mobile						
Tonga	0.2	13	22.7	36.0	403	13.91
Double drum	0.4	30	23.9	38.7	173	5.96
Permanent						
Brazilian modified	8.3	137	34.5	55.1	90	3.11
Hot tail	0.5	22	33.3	47.6	149	5.14
Mud beehive	7.2	38	32.0	44.8	101	3.49
Brick beehive 1	8.3	218	39.6	60.6	82	2.83
Brick beehive 2,3	2.0	105	37.5	62.5	95	3.26

a All costs are in 1989 $. Original cost figures have been converted to $ using a factor of 23 Baht to 1 US$ in 1983, and then inflated at 4 per cent per year for 6 years.

b
$$\text{Yield} = \frac{\text{weight of lump charcoal output (freshly recovered)}}{\text{oven dry weight of wood - weight of brands}} \times 100$$

The brands result from the incomplete conversion of wood to charcoal.

c Efficiency is the ratio of energy content in the charcoal output to the wood consumed.

Source: Compiled from Leach and Gowan, 1987.

Table 9.3
Charcoal Prices in Selected Urban Areas

City	Year	Price/Ton[a]	1989 GJ[b]	Source
Khartoum	1987	$175-220	6.53-8.21	Siemons et al., 1989
Nairobi	1985	$106-187	4.28-7.54	Hyman, 1986
Raipur	1985	Rs 1,590	4.23	Dunkerly et al., 1990
Hyderabad	1985	Rs 1,380	3.67	Dunkerly et al., 1990
Bangalore	1985	Rs 1,170	3.11	Dunkerly et al., 1990
Port au Prince	1989	$140	4.83	Destine, pers. com.

a Price per ton corresponds to the year noted on the previous column. $ values have been inflated to 1989 at the rate of 4 per cent per year; Indian rupees have been inflated at 7 per cent per year and converted to $ at Rs 17 per $.

b A higher heating value of 29 MJ/kg for air-dried charcoal has been assumed in calculating $/GJ.

Figure 9.5

A Traditional Charcoal Stove, Made of Metal

Photo credit: N. H. Ravindranath

Figure 9.6

Kenyan Jiko: An Improved Charcoal Stove where a Ceramic Liner and an Insulation Layer Placed within the Metal Shell Reduces Heat Losses

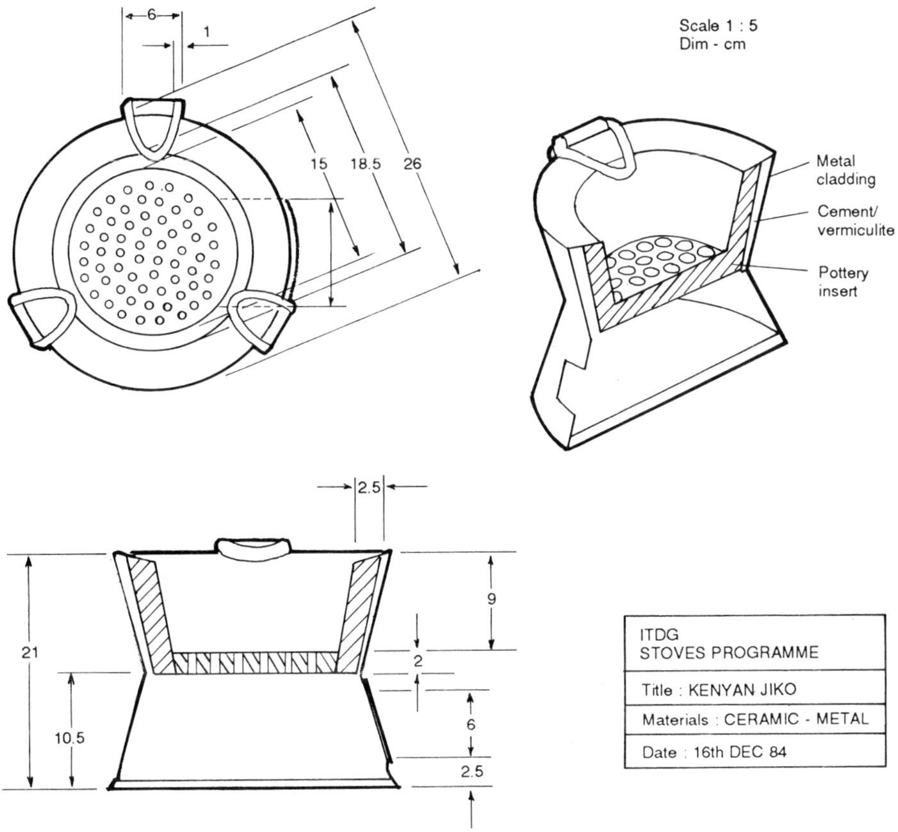

Figure credit: Intermediate Technology Development Group

Figure 9.7

Umeme - An All-metal Charcoal Stove Developed by UNICEF in East Africa

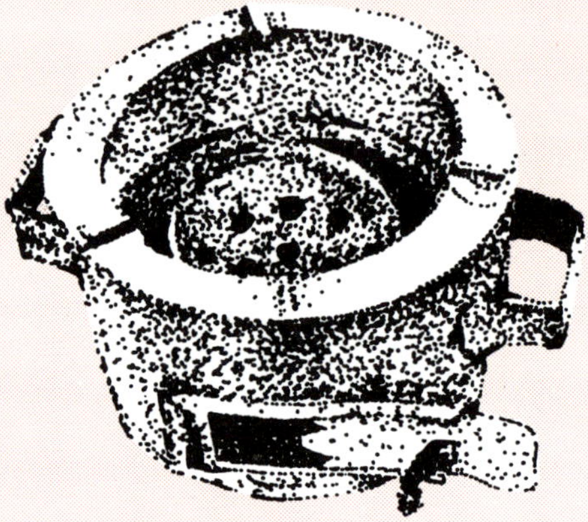

Photo: G. Dutt

Table 9.4
Some Performance Characteristics of Charcoal Cookstoves

Stove	Country	Air Control	Power (kW) High	Power (kW) Low	Efficiency High (%)	CO/CO_2 High	CO/CO_2 Low	Time to Boil
DUB 9	Burundi	y[a]	2.0	0.9	43.1	0.12	0.10	40
Haiti improved	Haiti	y	2.0	0.6	45.6	0.12	0.15	35
Ethiopian traditional	Ethiopia	y	0.9	–	43.0	0.12	–	75
CEPPE	Ethiopia	n	2.5	–	45.1	0.10	–	25
Sakkanal	Senegal	y	2.4	1.5	29.3	0.16	0.13	60
Feu Malgache	Sahel	n	1.8	–	29.2	0.12	–	65
Coalpot, improved[b]	Ghana	y	3.5	1.2	25.0	—	–	25
Sudan, traditional	Sudan	n	1.7	–	41.5	0.14	–	55
Thai bucket, traditional	Thailand	y	3.6	2.0	45.0	0.10	0.06	25
Thai bucket, improved	Thailand	y	4.0	0.8	48.6	0.04	0.06	25
UNICEF *jiko*[c]	Kenya	y	3.5	1.1	37.1	0.05	0.09	30
KENGO *jiko*	Kenya	y	2.8	1.6	45.2	0.08	0.07	30

a Where there was an air control, the stove was tested at both high and low power. The high-power test refers to heating up and boiling for a half hour. For the low-power test, water was simmered for an hour with reduced air opening. Low power efficiency was recorded but is not presented here.
b No gas analysis done when this stove was tested.
c This stove was also tested with an adapter for use with large pans (data not reported here).
Source: Sangen and Visser, 1986. Additional stove description (material, weight, and dimensions) are not recorded here.

a vertical cylindrical opening through the middle connected to a horizontal opening at the bottom. The fire burns in the cylindrical opening and heats the pan which is seated atop the can. Air is supplied from the opening below, through which a lit stick of wood is used to ignite the stove. The stove requires little or no attention, but the heat output cannot be controlled, the efficiency is comparable to that of an open fire, and it is very smoky. One prototype currently being developed provides secondary air to help burn the volatiles. It generates very little smoke and its efficiency is considerably improved, though still below that of the most efficient wood stoves (S. Dasappa, Dept. of Aerospace Engineering, Indian Institute of Science, Bangalore, pers. com. 1990).

Densified biomass. Since powdery biomass and many crop residues are loose and have low bulk density, they are difficult to transport. Consequently, many powdery agricultural residues that are unsuitable for fodder or fertilizer are not used as fuel either. One way to improve the volumetric energy density of the fuel is through densification or briquetting. The product has uniform dimensions and properties.

There are two basic ways of briquetting—with or without carbonization. Both require densifying but in one case the biomass is charred to form a charcoal-like fuel, while in the other no charring is involved. Briquetting machines are expensive. A small, manually operated press costs a few thousand dollars, while total investments for an 800-ton per year carbonizing plant were $90,000 in the Sudan (Siemens, Ahmed and Hood, 1989). Packaging, storage, and transport add to the cost of the final product. Cotton stalk briquettes (densified without carbonization) delivered to Addis Ababa were $114/ton in 1986—equivalent to $128 in 1989, assuming a 4 per cent increase per annum (Leitman, 1987). Considering an energy content of 17.8 MJ/kg (Leach and Gowan, 1987), the 1989 price is equivalent to $7.19/GJ.

Because of the large capital cost component, briquette prices are likely to be comparable in other countries. Many less expensive briquetting machines developed in India have run into serious technical problems. Nevertheless, briquetted biomass fuels were available in India at $40-60 per ton in early 1990 (EMC, 1990). Another estimate places briquetting cost of (1990) Rs 300 ($17) per ton excluding the cost of raw material (H.S. Mukunda, Indian Institute of Science, pers. com., 1991). Considering a raw material cost of half that of plantation-grown wood (1990 Rs 1/kg), the total cost would be about $72 per ton. This is substantially less expensive than in Addis Ababa. If we consider that in traditional stoves or open fires, densified biomass briquettes are less efficient than wood (Louvel, 1987), it is clear that briquettes must be cheaper than fuelwood for them to penetrate the market (Boiling Point, 1989).

Carbonization results in a higher quality fuel. Partially pyrolyzed briquettes produced one-thirtieth as much carbon monoxide as densified briquettes. Compared to charcoal, however, even carbonized briquettes are more difficult to ignite, and they generate more smoke. Moreover, except in a few places where charcoal is particularly expensive, biomass briquettes are likely to be more expensive on a $/GJ basis. To counter these problems, the Intermediate Technology Development Group developed briquette-based cooking systems consisting of both briquetting machines and special stoves (ITDG, 1985 and 1984; Young, 1988). Lower cost briquetting machines are also sold in Thailand.

Biogas. Biogas is produced by the anaerobic digestion of animal dung and other biomass. The gas can be burnt in a specially-designed stove that produces little carbon monoxide and no smoke. Furthermore, the digester slurry provides more fixed nitrogen to the soil than if the dung were directly applied.

Biogas digesters have been extensively promoted in some countries, with 5 million plants in China by 1986 and over a million in India as of 1988 (Qiu et al., 1990). A systematic survey of 4,108 plants built by 1984 in India found only 35 per cent working. Of the nonfunctioning plants, only 3 per cent were out of commission for technical reasons; 29 per cent did not have enough dung to operate; 16 per cent could not use the dung correctly; and 52 per cent were idle due to "lethargy, lack of interest, lack of knowledge about biogas, etc." (Dhawan, et al., 1990).

Most Indian biogas plants are small, intended for individual families with cattle. Economies of scale are possible with community and institutional biogas plants of which 430 had been installed in India by 1988 (TERI, 1989). Table 9.5 shows the costs of biogas plants as a function of production capacity. Installing piping to take the gas to village households and billing meters would considerably increase the capital investment required. Billing and collecting similarly could be expected to add substantially to the cost of operating and maintaining the digester alone. Preliminary estimates for these added costs, presented in Table 9.6, indicate a price of biogas delivered to households in the range of $11 to 12 per GJ.

Biogas can be produced from sources other than animal dung. A gas equivalent to biogas can be recovered from sewage treatment plants and put to some use. Similarly, effluent from certain food processing industries is an excellent raw material for biodigestion. The added advantage is that biodigestion considerably reduces the water pollution potential of these residues. More recently, nongrass weeds, leaf biomass, and inedible starch have received considerable attention as input for biodigesters. Moreover, technical improvements could increase mean residence time within a given digester volume (e.g., plug flow) or alter the process so that the residence time (and digester volume and cost) can be considerably reduced.

Burners specially designed for biogas are as efficient as natural gas or LPG burners, i.e., around 60 per cent (Jiang et al., 1984). In one study of six Indian biogas stoves, however, efficiencies varied from 40 to 65 per cent (Vasudevan and Jain, 1989). Variation in gas supply pressure, one factor that reduces efficiency, may be reduced by incorporating a pressure regulator into the digester design.

Producer gas. Gaseous fuel may also be produced from biomass by thermo-chemical conversion. Raw biomass is partially oxidized to produce a combustible gas, primarily a mixture of carbon monoxide and hydrogen. In many industrialized countries, coal-derived producer gas was supplied to households for cooking and lighting before natural gas became more widely available. In Beijing and Shanghai, about half of all households still use it for cooking (Zhu et

Table 9.5
Costs of Biogas Plants in India[a]
(Plant life 25 years)

	Capacity m^3/day	Digester[b] (1990 $)	O & M[c] (1990 $/yr)
Individual	2	369	9
Community	12	883	399
	24	1,322	548
	30	1,541	671
	36	1,760	745
	45	2,097	908
	54	2,418	1,113
	60	2,638	1,187

a Original data in 1985 and 1986 Rs.
b Additional capital costs for a biogas plant include piping and connections of the digester, sand filters for the slurry output of the digester, and tools and accessories for maintenance.
c For community scale plants, the operation and maintenance costs include fuel cost, equivalent to gathering dung and distributing the slurry produced by the digester.
Source: Larson, 1991, citing Reddy 1991 and Bhatia 1990.

Table 9.6
System Cost Estimate for a Community Biogas Plant
Supplying Cooking Gas to 40 Households[a]

Capital investment	**US$**	
Biogas digester, 45 m^3/day	2,100	(rounded, from Table 9.5)
Distribution piping	2,100	(est.)
Meters, 40 @ $50	2,000	(est.)
Total	**6,200**	
Annual operation	**$/yr**	
Operation and maintenance of digester	900	(rounded, from Table 9.5)
Billing and administering	1,800	(est.)
Total	**2,700**	

Life of biogas plant = 25 years

Total gas demand = 40 households x 6 persons/household x 0.15 ft^3/day x 22.3 MJ/m^3
= 802.8 MJ/day or 293 GJ

Discount rate	0.06	0.12
Annualized investment and operation	$3,185	$3,491
Gas cost delivered, $/GJ	11	12

a We assume a 6-person household, with a per capita fuel consumption of 0.15 ft3/day as measured in the Ungra experiments. This gives an average gas demand of 36 ft^3 (1m^3) per day. A 45 ft^3 (1.3 m^3)/day capacity biogas digester is selected. The calorific value of gas is shown in Table 9.1. A number of costs have been estimated and are indicative of the expected magnitude.

al., 1973). In a number of urban areas in developing countries, too, a coal-based producer gas remains in operation. A charcoal-based producer gas was used as a transport fuel in Sweden and elsewhere during the Second World War, when petroleum availability was limited (SERI, 1979). Recently, biomass gasifiers have been proposed and developed to fuel motors in a number of petroleum-importing developing countries (Dasappa et al., 1985; Kjellstrom, 1985; and Bernardo and Kilayko, 1990). Cookstoves using producer gas exist in China, in Calcutta (where it was available until recently), and presumably elsewhere with a history of town gas use.

In recent years, producer gas has been proposed and used principally for motive power (pumping, etc.) and for electricity generation in rural areas. Costs of Indian gasifiers from two sources are given in Table 9.7. In these uses the gasifier equipment includes gas cleaning and cooling, which is necessary before the gas can be fed to an internal combustion engine. These are not needed for cooking use. Instead, a blower is needed to draw gases through the gasifier, and a gas storage tank. To estimate cost, we can assume that the price of a cooking gasifier system is the same as that for an engine (though it will probably be less). As in the case of biogas, there will be additional investment in distribution piping, metering, billing and administration, as well as costs for maintenance and operation. Table 9.8 shows an estimate of overall costs of a producer-gas based cooking system, and indicates that the price of gas delivered to households is about $15-16/GJ.

While there are significant economies of scale in gasifier costs, there are relatively little economies in distribution and metering costs.

Table 9.7
Costs of Wood Gasifiers, Including Cooling and Cleaning
(for use in engines)

Source	**Capacity (GJ/h)**	**Capital Cost ($)**
Jain	0.0876	1,385
	0.164	2,077
	0.385	4,377
	1.425	15,385
Baliga	0.0855	1,389
	1.44	30,556

Sources: Jain and Vyarawalla, 1985, and Baliga et al., 1990. Original figures have been converted to 1990 $ using a conversion factor of Rs (1985) 13 = $1 (1990).

Table 9.8
System Cost Estimate for a Wood Gasifier Supplying Cooking Gas to 40 Households[a]

Capital investment	**$**		
Gasifier, 0.164 GJ/h	2,100	(rounded, from Table 9.7)	
Distribution piping	2,100	(est., same as Table 9.6)	
Meters, 40 @ $50	2,000	(est., same as Table 9.6)	
Total	**6,200**		
Annual operation	**$/yr**		
Operation and maintenance	620	(10% of investment)	
Fuelwood cost	1,460[b]		
Billing and administering	1,800	(est/. same as Table 9.6)	
Total	**3,880**		
Life of gasifier = 25,000 hours = about 10 years at 7 hours/day.			
Life, balance of system = 25 years.			
Discount rate	0.06		0.12
Annualized investment and operation	$4,486		$4,774
Gas cost delivered, $/GJ	15		16

a We assume a 6-person household with an annual cooking energy demand of 293 GJ, as in Table 9.6. We assume a 70 per cent efficient gasifier; actual efficiency is likely to be higher, but we have not included here the (primary) energy consumption or cost of the electric blower operation. Thus fuelwood input will be 419 GJ/yr. For a 0.164 GJ/h (input) gasifier this implies a 7-hour a day operation, a reasonable work day. Increased demand could be met, provided wood was available, by operating the same gasifier for longer hours.

b For fuelwood input of 419 GJ and fuel price of $3.49/GJ, we get $1,462.

Moreover, Table 9.8 shows that the costs are dominated by fuel, operation, maintenance, billing, and administration.

Ethanol. Kerosene is an important cooking fuel among the urban poor in many developing countries. Some oil-importing countries could use alcohol as a cooking fuel. Alcohol is used to enhance gasoline octane in a number of countries and to fuel much of Brazil's automotive fleet.

At least two alcohols, methanol and ethanol, can be produced from biomass. Ethanol can be made by fermenting sugarcane, sweet sorghum, cassava, sweet potato, maize, etc. and from wood and other cellulosic materials (Baliga et al; 1990). Methanol, also called wood alcohol, was formerly made from wood by a thermochemical process; at present it is made primarily from natural gas.

Brazil has the most experience in the use of ethanol, derived from sugarcane as fuel, for automotive uses. A number of studies estimated the costs of ethanol, as shown in Table 9.9. They differ somewhat, but we may consider $(1990)11/GJ to be a typical estimate. The cost of delivered cane is the single most important determinant of the cost of ethanol (Larson, 1991). Alcohol derived from sugarcane has been proposed as a cooking fuel for Mauritius, where alcohol cookstoves have also been tested (Bagvant and Panray, 1991). For a plant with an annual output of 18 million liters of low grade ethanol (85 per cent alcohol), Bagvant and Panray estimated a production cost of $0.20/liter (1990) in Mauritius. At a calorific value of 16.7 MJ/liter (for 85 per cent ethanol), this works out to $12/GJ, slightly higher than the Brazilian typical estimate. One way of reducing alcohol production costs is through an integrated cane sugar factory and alcohol distillery, where sugarcane residue (bagasse and barbojo) is used for cogeneration with biomass gasifiers and gas turbines, a technology under development (Ogden et al., 1990). The price of ethanol depends on the technology, duration of the milling season, and how the sugarcane residue and electricity are priced. We assume a figure of $0.17/liter, equivalent to $7.08/GJ, considering 24 MJ/liter of ethanol (Larson, 1990).

Table 9.9
Estimates of the Cost of Ethanol Production in Brazil

Source	1990 US cents per liter	$/GJ
World Bank, 1984	22.5 – 24.3	10.2 – 11.1
Borges, 1984	20.1 – 27.4	9.1 – 12.5
CENAL, 1984	25.3 – 28.2	11.5 – 12.8
Mello & Pelin, 1984	49 – 56	22.0 – 25.0
Comissao Nacional de Energia, 1987	21.3 – 23.1	9.7 – 10.5
Motta, 1987	28.0 – 44.0	12.7 – 20.0

Source: Copersucar 1989 as cited in Larson, 1991.

In many countries, land cannot be diverted from food production to grow sugarcane for alcohol. However, other routes to alcohol production do not involve a food-fuel conflict. For instance, ethanol can be fermented from the nonedible stems of sweet sorghum. Field tests have led the Nimbkar Agricultural Research Institute to propose the growing of sweet sorghum for three products—grain for human consumption, alcohol from the stems, and bagasse for fodder or energy (Rajvanshi, 1988).

In India, 17.6 million hectares were cultivated to sorghum in 1979 (Rajvanshi, 1988). If this area were converted to sweet sorghum cultivation, the alcohol yield would be 18 million m^3 per year, the energy equivalent of 9 million tons of kerosene.

One government estimate places kerosene demand in India for the year 2000 at 18 million tons (ABE, 1985). Thus, sweet sorghum could contribute substantially to reducing the demand for imported kerosene without reducing grain yield. The grain yield of sweet sorghum is higher than the normal variety and far higher yields have been observed than were used in making this estimate of alcohol potential (Rajvanshi, 1988). On the other hand, sweet sorghum does not taste the same as normal sorghum and there may not be enough demand for the sweet variety to permit a major shift in sorghum plantation.

Alcohol fuels are not commonly used for cooking. A number of camping stoves use ethanol or denatured alcohol. Some of the simpler, nonpressure models should be adaptable to cooking in less-developed countries. Gill (1986) studied some camping stoves and measured efficiency in water-boiling tests of around 60 per cent. Baguant and Panray (1991) also found an efficiency of 60 per cent in one stove they tested. The Vidyarthis (1985) found efficiencies comparable to kerosene stoves. A program of testing liquid and gas-fuel cookstoves conducted by Eindhoven University of Technology for the World Bank provided insight on factors that would affect the performance of alcohol stoves as well (EUT, 1985). The Nimbkar Agricultural Research Centre has also developed an alcohol cooking stove (Rajvanshi, 1988 and 1984).

Fuel Choice and Comparative Cooking Fuel Economy

Cooking systems may be compared on the basis of fuel preferences, relative energy consumption, and comparative costs.

User fuel preferences. Table 9.10 shows a hierarchy of fuel preferences. Findings are based on surveys and on how fuel choice changes with increasing household income. The left column is based on observations in urban areas, where many fuels may be simultaneously available. An increase in income is associated with a rise up the fuel-preference ladder (Leach and Mearns, 1988). The right column notes where some alternative fuels might be located relative to the more common fuels.

At point of use, electricity is the cleanest "fuel." However, because of its expense it is rarely used for cooking in less-developed countries. Natural gas is a convenient fuel because, like electricity, it

Table 9.10
User Preferences for Cooking Fuels
(listed in decreasing order of preference)

Common Fuels	Alternative Fuels
Electricity	
Natural Gas	
Liquified petroleum gas	
	Biogas (centralized plants)
Kerosene	Producer gas
	Alcohol
Charcoal	
Coal and soft coke	
	Coal briquette
	Biogas (individual plants)
	Densified, carbonized biomass briquette
Fuelwood	
	Densified biomass briquette
Crop Residue	
Dung	

is distributed directly to the home. Some societies (e.g., France and Costa Rica) have shown significant concern about fire and explosions from gaseous cooking fuels.

LPG has a slight disadvantage in that gas cylinders take up space and have to be replaced when supply runs out. (In some locations, the homes have fixed tanks that are periodically filled by a tanker truck.) In some countries, and particularly in rural areas, LPG delivery may be uncertain.

Kerosene is a smelly fuel that blackens pots. Kerosene stoves are not as convenient to light and put out as gas stoves. Thus, kerosene ranks below gaseous fuels but above solid fuels.

Among solid fuels, charcoal ranks highest, and was the premier cooking fuel until the recent arrival of petroleum fuels. It is less easy to light and extinguish than gas or liquid fuels. In common charcoal stoves, heat output cannot be readily controlled. Coal, soft coke, and coal briquettes have similar characteristics.

Firewood is the primordial fuel. Even the most modern societies have been reluctant to part with burning wood in their fireplaces, although in centrally heated houses, there is little thermal benefit. In urban Java, Fitzgerald, Barnes and McGranahan (1989) observed that even the wealthy used fuelwood for cooking if it is available locally.

Crop residues and dung tend to be fuels of last resort and are used principally in regions where fuelwood is in short supply. Noncereal crop residues are used extensively as cooking fuel in South and Southeast Asia, where cereal residue is used as fodder. In China, straw is the principal cooking fuel in rural areas (Qiu et al., 1990).

Though not all fuels mentioned above are available everywhere, relative preferences are consistent with this hierarchy. Among the alternative fuels, biogas, like natural gas, is usually piped to the kitchen. On the other hand, the heating value is substantially lower than that of LPG. Thus, biogas provided to households from a centralized plant might rank between LPG and kerosene.

Producer gas derived from biomass has not been extensively used as a cooking fuel. The gas contains carbon monoxide and there is some danger of carbon monoxide poisoning from leaking fixtures. We tentatively rank it at par with kerosene. Ethanol is almost never used as a cooking fuel. As a liquid fuel, it should rank with kerosene but, because of lower heating value and possible problems (e.g. mixing water), we place it below kerosene. In societies/where alcohol production is illegal, it is unlikely to be an alternative cooking fuel. Briquetted biomass fuels are ranked as shown for reasons explained above (see Stoves for Biomass Residues).

Energy use with different fuels. In the above discussion, we considered fuel preferences without regard to energy use or economics, although these would affect how people use different energy forms for cooking. Dunkerley et al. (1990) analyzed energy used by households in several Indian cities. Fuelwood consumption decreased with increasing income since middle-and upper-income households tended to use kerosene and LPG. The total amount of cooking energy use was independent of income. (Fitzgerald [1989] noted a similar result in urban Java.) Assuming efficiencies of conversion for the various fuels, Dunkerley et al. (1990) showed that the amount of cooking increases rapidly with income, more than doubling over the income range considered. They also noted differences across cities, in part resulting from the availability and prices of alternative cooking fuels. In rural areas, where alternative cooking fuels may not be available, increased cooking with increased income is reflected in greater fuelwood use.

To account for income and other factors, fuel substitution ratios can be estimated from a regression of energy used for cooking, and adjusted for household income, household size, fuel price, prices of competing fuels, an index of fuel availability, cooking practices, and combinations of fuels used. Fitzgerald, Barnes and McGranahan (1989) used this method for urban Java to compare energy use for cooking with fuelwood, kerosene, and LPG. They found that households cooking with fuelwood consume 1.66 times as much energy as those cooking with kerosene, while those cooking with LPG consume 0.85 times as much. Even assuming an average wood stove efficiency as high as 20 per cent and a kerosene stove efficiency of 40 per cent, fuelwood users should consume more on the basis of relative efficiencies. Similarly, LPG users consume more relative to kerosene users than would be suggested by their respective efficiencies. The conclusion is somewhat surprising: even adjusted for socioeconomic factors, energy savings are less than would be expected from technical efficiency considerations as households "upgrade" from fuelwood to kerosene to LPG. This must be taken into account in programs to introduce upgraded cooking fuels.

Savings "loss" can also be expected relative to efficiency change when an improved stove using the same fuel is introduced. Where fuelwood or charcoal is commercially purchased, any savings realized from a fuel-efficient stove reduces fuel expenditure. This has the effect of increasing the disposable income of the households which clearly is a desirable goal. But if the goal of an improved stoves program is also to reduce fuelwood consumption, then we should determine the extent to which households "take back" some of the savings by cooking more and spending the money on other items that include fuelwood as an input. The actual savings in fuelwood from an efficient stove are therefore less than would be expected. The actual savings can be determined by measuring fuel use before and at various intervals after the introduction of an improved stove. Ravindranath et al. (1989) reported a sequence of such evaluations of the ASTRA ole. Jones (1988) showed that additional savings "leakage" occurs if a fuel is used principally for cooking (e.g., charcoal) and fuel price responds to reduced demand from an improved stove.

Energy requirements with different cooking systems. Some of the above conclusions depend on comparisons based on the relative efficiency of different cooking systems. The researchers assumed efficiency figures for different cooking fuels (without mentioning the type of stoves in use). Other studies calculated the relative economics of alternative cooking systems based on assumed efficiencies (Kishore, 1988; Ultria, 1989). Based on WBTs, efficiencies of the same stove vary widely with the test circumstances (Gill, 1987; Bussman, 1988).

One comparison of the stoves' technical performance may be through controlled cooking tests. In Fiji, Siwatibau (1981) compared performances of two kerosene, a charcoal, and three fuelwood stoves, each used to cook five different typical meals representative of the islands' Fijian, Indian, and Chinese populations. The specific fuel consumption—energy input per kg of cooked food—based on the average of the five meals is shown in Figure 9.8.

Figure 9.8

Specific Energy Consumption for Various Stove/Fuel Combinations in Fiji, Averaged over Five Different Meals Typical of Chinese, Indian and Fijian Households

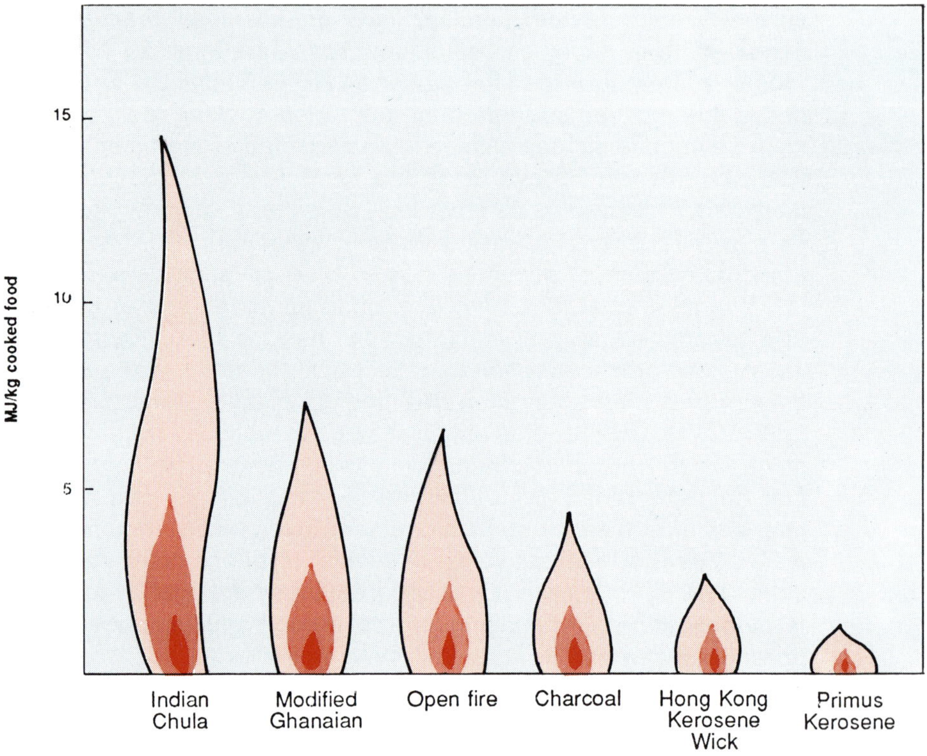

Source: Data based on Siwibatu 1981

The data show a large range of energy use, exceeding a factor of 11 from the lowest (Swedish primus, a kerosene pressure stove) to the highest (Indian *chula,* a four-burner cement fuelwood stove with a chimney). Notably, the latter consumes well over twice the fuel of an open fire, which turns out to be the most efficient of the wood stoves. The kerosene wick ("Hong Kong") stove uses more than twice the fuel as the pressure stove, and shows large variations in the use of high quality fuels. The less efficient kerosene stove uses about 2.5 times less energy than the open fire. The Fiji study did not compare energy-efficient wood stoves with other options. EUT (1985) gives an interesting account of using WBT data to predict fuel use in these cooking trials.

A comparison of stove and fuel combinations in actual cooking was also conducted in December 1990 in Ungra Village, Karnataka. A typical village meal for a 6-person household (rice, millet, cowpeas, and vegetables) was cooked three times in each of nine stove/fuel combinations. Fuels used were firewood, charcoal, dung cake, biogas, kerosene, LPG, and electricity. Stoves used are described in the footnotes to Table 9.11. A village cook familiar with several of the stoves did the cooking in all except the electric and biogas stoves. For technical reasons, these two could not be tested in the village. The fuel consumption data recorded in these measurements are summarized in Table 9.12.

At the point of use, electricity has the lowest specific energy consumption. If we include efficiencies of electricity generation (35 per cent) and transmission and distribution (optimistically 85 per cent), the primary energy consumption with electricity would be 0.64/(0.35 x 0.85) = 2.15 MJ/kg, only marginally less than the most efficient fuelwood option (2.24 MJ/kg). The most efficient fuelwood option uses 1.82 and 2.47 times the least (biogas) and most (LPG) efficient of the liquid or gas-fueled systems. The traditional wood stove consumes 3.76 as much as the Perfect kerosene stove, and the LPG stove consumes 0.79 times more.

The adjusted fuel-substitution ratios estimated for urban Java were 1.66 for wood/kerosene and 0.85 for LPG/kerosene. If the Ungra data on relative efficiencies are anywhere close to the values in Java, the data confirm a substantial leakage of energy savings in switching from wood to kerosene. (Controlled cooking tests should be conducted in Java to determine a more precise measure of the leakage. Also, the Java-type regression analysis should be conducted with survey data and controlled cooking comparisons in other regions to estimate savings leakage in fuel switching.)

Table 9.11
Comparative Economics of Fuel/Stove Combinations for Cooking

Fuel	Stove	Stove Cost[a] US$	Stove Life[b] Years	Annual Fuel Use[c] GJ	Fuel Cost $/GJ	Annualized Lifecycle Cost, $ at Real Discount Rates of 0.06	0.12	0.75	1.50
Part A: Based on Cooking Fuel Use in Ungra Village (from Table 9.12)									
Wood	3-stone	.00	3	20.27	3.49[d]	71	71	71	71
	Traditional[e]	.00	3	25.12	3.49	88	88	88	88
	ASTRA	7.78	3	13.64	3.49	51	51	55	60
	Swosthee[f]	5.56	3	17.15	3.49	62	62	65	69
Charcoal	traditional	3.00[g]	1	14.49	8.86	132	132	134	135
Dung cake	traditional	0.00	3	24.35	note	–	–	–	–
Sawdust	IISc[j]	(5.00)	2	24.35	1.74	45	46	48	52
Biogas[k]	Individual	13.89[l]	7	7.33	0.00	38	56	286	563
	Community	13.89	7	7.33	(11.50)	87	87	95	105
Kerosene	Nutan[m]	6.94	3	6.35	5.30	36	37	40	45
Perfect[n]		8.06	3	6.78	5.30	39	39	43	49
LPG	Superflame[o]	50.00	7	5.40	5.30	49[p]	53	107	180
Electricity	Hot plate[q]	38.89	5	3.84	25.00	105	107	127	155
Part B: Estimates, Not Based on Measurements									
Charcoal									
Improved *jiko*		5.49[r]	1	(11.00)	8.86	103	104	107	111
Briquette[s]		(5.56)	3	17.15	(7.19)	125	126	128	132
Producer gas[t]		(13.89)	7	7.33	(15.50)	116	117	124	134
Alcohol[u]									
Brazil conv.		(8.00)	3	6.78	11.00[v]	76	76	81	87
Advanced technology		(8.00)	3	6.78	(7.08)[w]	51	51	55	61

a Stove costs are market prices in Bangalore; Swosthee is not commercially available—the price is an estimate based on the production run of an earlier version (H.S. Mukunda, U. Shrinivasa, and S. Dassappa, 1988). A conversion rate of Indian Rs. 18 per US dollar is used, corresponding to the official exchange rate in December 1990. Stove costs are annualized at the discount rates shown.

b Kerosene stove life has been adjusted down, partly since the replacement cost of wicks is not considered. Stove maintenance costs are neglected for all of the options.

c Annual fuel use is estimated from measured energy use per meal, by considering two identical meals a day, as is the custom in the Ungra village region and corresponds to a family of six.

d Fuelwood cost is taken to be $0.0555/kg equivalent to the cost of plantation wood in India of about Rs 1/kg. Charcoal cost is estimated at 2.54 times that of plantation fuelwood in Bangalore in the early 1980s—$8.86/GJ (Dunkerley et al., 1990, Reddy and Reddy, 1983). Fuel price of sawdust, often a "waste" material, is taken to be half that of fuelwood. For biogas costs, see Note k below. Cost of kerosene and LPG is assumed equivalent to $30 per barrel of petroleum (including refining and distribution); electricity has been assumed to cost $0.09/kWh corresponding to a marginal cost of supplying electricity from new power plants.

e The traditional stove is a three-pan stove without chimney (Fig. 9.1); the ASTRA stove is a three-pan stove with chimney, designed to be energy efficient (Fig. 9.2).

f Swosthee is a single-pan metal stove designed at the Institute of Science. The version MS-4 (Fig. 9.5) was tested here.

g The charcoal stove was a traditional metal stove, similar to the West African Feu Malgache and the Kenyan jiko, which costs Kenyan shillings (KS) 35-45 in 1985 (E.L. Hyman, 1986); we assume a cost of KS 40 ($2.92 in 1989). We use the Kenyan price to facilitate comparison with the Kenyan "improved jiko" shown in part B of the table. The parentheses here and elsewhere in the table indicate an estimated value. The improved jiko (Fig. 9.8) sold for 60-85 KS in Nairobi in mid-1985 and with a saturated market, a retail price of KS 55-75 was expected (Hyman, 1986). We take a price of KS 75 ($5.49 in 1989). Traditional stoves last about a year, while the various components of the improved stoves last from six months to three years. We assume a one-year average lifetime for both. We would expect cooking energy consumption with the improved stove to fall between an efficient wood stove and a kerosene stove.

h Dung cake was burnt in the traditional three-pan stove normally used with wood. When dung cake is burnt, this stove is used.

i For dung cake we should take the value of an equivalent amount of fuel as its opportunity cost. This figure was not at hand.

j This sawdust stove is similar to a traditional stove modified to improve combustion by providing secondary air. It was a prototype developed at the Indian Institute of Science. (See above, Powdery Biomass)

k For biogas, two cases are considered here:
"Individual" is a 2 m^3/day family-sized digester, with cost data as shown in Table 9.5. The digester is owned by the household and capitalized at its discount rate. The cost of family labor in animal husbandry and digester operation is not included.
"Community" is a community-scale biogas plant operated as a public utility that pipes and sells gas to households. Gas cost is the average of estimates for discount rates of 6 and 12 per cent.

l The biogas stove was a standard burner designed by the Khadi and Village Industries Commission—the agency that promotes biogas in India.

m The "Nutan" is a fuel-saving wick stove developed by Indian Oil Corporation.

n The "Perfect" is a popular brand of wick stove.

o The "Superflame" is a two-burner LPG stove.

p Annualized cost for LPG includes a $50 gas cylinder with an expected life of 5 years.

q For electricity, a single-burner hot plate was used.

Assumptions specific to Part B of the Table:

r We consider a metal stove equivalent to the Swosthee MS-4 with the same cost, lifetime, and fuel consumption rates as when it is used with fuelwood. We assume a fuel price as estimated on above, see Improved Fuelwood Stoves.

s We consider the case of a gasifier managed as a public utility that pipes the gaseous fuel to the household. The stove is assumed to cost the same as a single burner biogas stove, have the same life, and consume the same amount of fuel. We assume a unit cost of gas as estimated above, see ***Producer Gas.***

t We assume ethanol stove fuel consumption characteristics to be the same as a kerosene stove.

u Ethanol cost typical of Brazilian distilleries, see Table 9.9, and ***Ethanol***.

v Advanced technology with bagasse gasification and gas turbine cogeneration, see ***Ethanol***.

Table 9.12
Efficiency and Fuel Consumption of Stove/Fuel Combinations (Based on Water Boiling and Controlled Cooking Tests in Ungra, Karnataka, India)

Fuel	Stove	Efficiency[a] (WBT, %)	Specific Fuel Consumption[b] Physical units/kg		MJ/kg	Average Cooking time (mins.)
Biofuel Stoves[c]						
Wood	3-stone fire	15.6	217	g	3.44	101
	Traditional 3-pan	14.2	271	g	4.31	62
	ASTRA ole 3-pan	33.2	141	g	2.24	62
	Swosthee MS4	17.2	183	g	2.91	111
Charcoal	traditional metal	32.2	95	g	2.38	N.A.
Dung Cake	traditional 3-pan	11.1	304	g	4.00	N.A.
Sawdust	IISc improved	30.4	253	g	4.02	N.A.
Biogas	KVIC burner	45.1	0.05	m^3	1.13	106
Other Stoves						
Kerosene	Nutan	60.2	26.1	g	1.13	106
	Perfect	40.4	26.6	g	1.15	131
LPG	Superflame	60.4	20.1	g	0.91	76
Electricity	hot plate	71.3	0.17	kWh	0.64	99

a The efficiency is the average of three water boiling tests, except for the three-stone fire and the ASTRA stove (2 tests each).
b Specific fuel consumption is based on kg of cooked food, averaged over three meals in each case. The typical meal 1,102 g rice (+2,121 g water); 883 g ragi flour (a millet, + 1,554 g water); 256 g cowpeas (a legume, + 2,463 g water); and 150 g vegetables mixed with the cowpeas.
c The stoves are described in the footnotes to Table 9.11.

The Ungra cooking trials were accompanied by a series of standardized WBTs – their average efficiencies are shown in Table 9.12. The relationship between specific fuel consumption in cooking and WBT efficiency is explored in Figure 9.9, which includes the range of measured values and the averages of Table 9.12. As expected, specific fuel consumption decreases as efficiency increases, but the relationship is weak in the fuelwood stoves and kerosene and gas stoves. For instance, the traditional wood and the three-stone fire stoves have virtually the same efficiency, but the traditional stove consumes 25 per cent more fuel. On the other hand, the two kerosene stoves have substantially different efficiencies (40 per cent and 60 per cent), but their fuel consumption in cooking is virtually the same. Similarly, the specific fuel consumption of the electric stove is substantially smaller compared to the wood stoves than would be suggested by a ratio of efficiencies alone. The discrepancies are presumably due to differences in low-power performance that are not

Figure 9.9

The Relationship Between Specific Fuel Consumption to Cook a Typical Ungra Village Meal and Efficiency in Water Boiling for Stove and Fuel Combinations

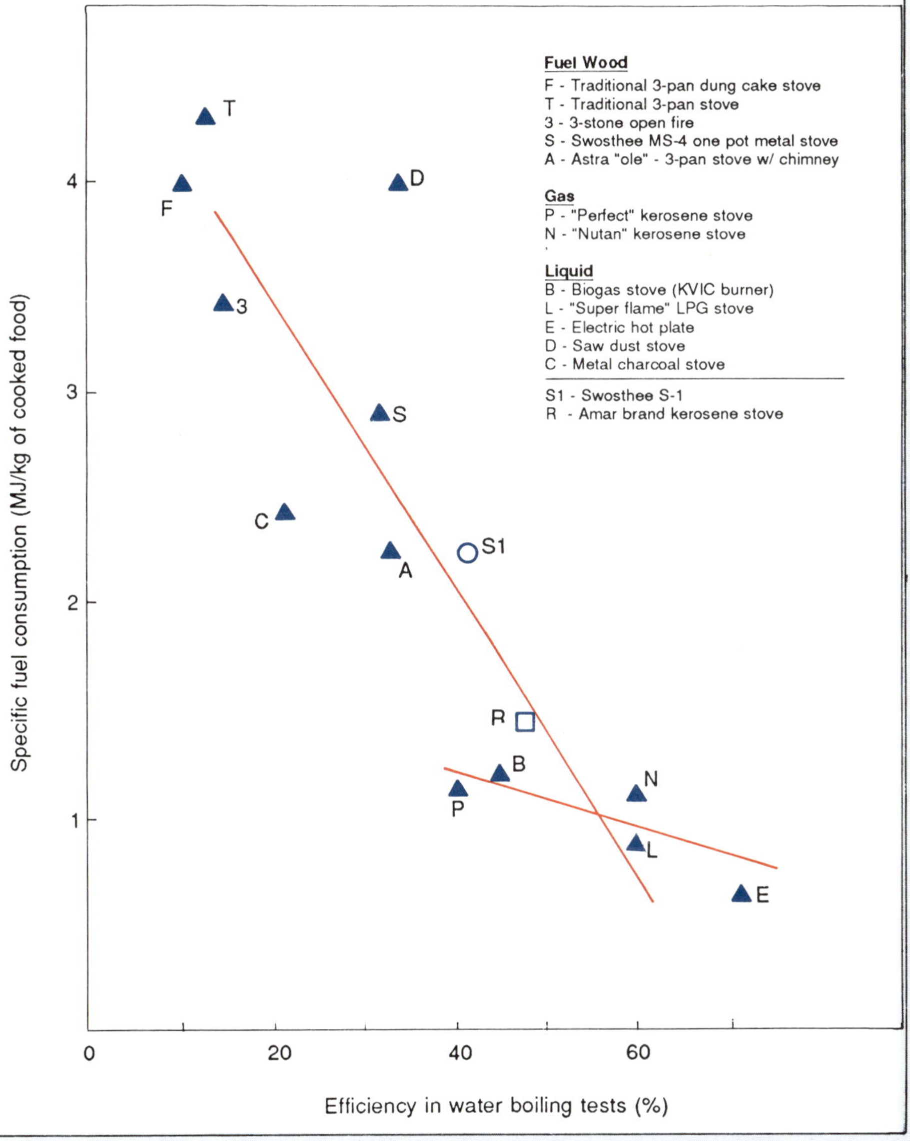

characterized in our simple measurement of water boiling efficiencies (EUT, 1985).

Table 9.12 also shows the average cooking times for the various options. The traditional three-pan and the ASTRA three-pan stoves are the fastest, while cooking on a (single) kerosene stove can take more than twice as long. Cooking time is an important consideration in stove choice.

Economics of alternative cooking systems. We have used cooking fuel consumption to make an economic comparison of stove/fuel alternatives, as shown in Table 9.11. Instead of actual fuel prices, we based estimates on the international prices of petroleum fuels, and marginal cost of electricity supply and cost of plantation fuelwood for India. Thus, we have sought a societal perspective although an environmental cost or a foreign exchange premium for oil-importing countries with balance-of-payments problems has not been included.

The cooking cost is calculated in terms of an annualized life-cycle cost that includes the cost of the stove and auxilliary equipment (gas cylinder for LPG and biogas plant for the biogas option), and their respective lifetimes. Capital costs are annualized using several discount rates. The values of 0.06 and 0.12 (real discount rates of 6 and 12 per cent) represent bounds on the cost of capital to society. The discount rates implicit in purchases that involve mixes of capital and operating expenses are considerably higher for low-income households, reaching 150 per cent for the poorest households (A.K.N. Reddy, pers. com., 1990). Thus, the economic comparison in Table 9.11 also shows annualized life cycle costs at real discount rates of 0.75 and 1.50 (75 and 150 per cent).

Note that the high implicit discount rates observed in consumer purchases are not necessarily the result of capital shortage. Das et al. (1990) estimated the information costs for learning about new technologies such as improved fuelwood stoves and solar cookers. If these information costs are included, the new technologies are not the cheapest options, and consumer behavior is indeed economically rational. Better promotion of lower cost options will reduce the private information cost (and "implicit" discount rates) and may permit new options to be accessible.

We first compare the options in Part A of Table 9.11 that are based on actual capital costs and fuel use as measured in the Ungra cooking trials. At all discount rates, the kerosene stoves are the least-cost option and electric cooking, not surprisingly, is one of the most expensive. Charcoal in a traditional stove turns out to be even more expensive than electricity, because the fuel cost is high and the fuel consumption is comparable to the efficient wood stoves. If sawdust is available at half the cost of fuelwood, the improved sawdust stove developed at the Indian Institute of Science would be the least expensive biofuel option, especially at high discount rates. More generally, the ASTRA stove is the least-cost biofuel option, especially at high discount rates. (Note that we assume a fuelwood cost based on the price of plantation wood, a surrogate for wood grown renewably, while many people in rural areas may be able to gather it at no monetary cost.)

At a low discount rate of 6 per cent, biogas from an individual digester is slightly cheaper, mostly because the labor for operating the plant is not included. Because it is capital-intensive, the individual biogas option increases in cost as the discount rate increases. In the community biogas case, distribution piping and operation, maintenance, and administrative costs are included, although all capital investments are calculated at social discount rates. The result is that the community biogas option is much more expensive at low discount rates. Though it is much less than the individual biogas case at high discount rates (since the plant capital expenses are still at the lower social discount rate), it is more expensive than most of the other options. At the social discount rates of 6 and 12 per cent, the LPG and the most efficient woodstove (ASTRA) options are virtually identical in cost; the high capital costs of the gas stove and the gas cylinder, however, make this alternative very expensive at the implicit discount rates of low-income households.

In Part B of Table 9.11, estimated costs and energy use include several biofuel alternatives that were not tested at Ungra—charcoal, sawdust, briquetted biomass, producer gas, and alcohol. While these figures are tentative, several observations can be made.

Among the gas and liquid fuel options, kerosene remains the least-cost option. Even with an improved stove, cooking with charcoal would remain expensive. Briquetted fuel would also be an expensive alternative. Producer gas is also likely to be expensive using plantation-grown wood. Using Brazilian data for alcohol production, cooking with alcohol would cost much less than the expensive alternatives (charcoal, electricity, producer gas, or briquetted biomass) but would still be twice the cost of kerosene and about 1.5 times that of the ASTRA stove.

The advanced technology alcohol option is comparable to the LPG option at low discount rates and virtually identical to the ASTRA ole at all discount rates. In a way this is an unfair comparison since we have not considered advanced technologies for the other cases. Social constraints rather than technology are likely to limit further fuel savings with wood stoves in rural areas. We have found that after some time, users do not operate the ASTRA stove as intended—they do not close the firewood bins—and thus they forego part of the savings (Ravindranath, Shailaja and Revankar, 1989). Research into gas- and liquid-fueled stoves suggests that additional savings may be possible through technological innovation. For instance, the water-boiling efficiency measured for the biogas burner was only 45 per cent, compared to 60 per cent with the LPG burner. As noted before, the efficiency of biogas burners can be greatly improved, matching that of LPG burners. Even the efficiency of a conventional LPG stove can be improved. An infrared impingement burner developed for use with natural gas was found to have efficiencies of 65-70 per cent, and much lower nitrogen oxide emissions than a conventional burner (Shukla and Murley, 1983). An analysis of the performance of wick-type kerosene stoves showed that fuel consumption could be significantly reduced even without increasing the water-boiling efficiency if the stoves could be designed to provide a wider range of power output (EUT, 1985). The efficiency of gas- and liquid-fueled stoves is less dependent on the operator, and technical improvements are likely to lead to savings in actual cooking as well.

We may also observe in Table 9.11 that where costs are dominated by fuel use and fuel cost rather than stove and associated equipment cost or lifetime, the annualized life-cycle cost is weakly dependent on discount rate. This is in fact the case for all except the LPG, electric, and biogas options. One consequence is that the results are generally not very sensitive to assumed stove cost or lifetime. This suggests that increased costs of more efficient stoves are likely to be cost-effective. For instance doubling the kerosene stove cost (to $13.88, say) while reducing the fuel consumption by 10 per cent relative to the Nutal stove reduces the annualized life-cycle cost for the lower discount rates.

Cooking, Resources, and the Environment

In the previous section we discussed user fuel preferences, energy use, and economics of alternative cooking systems without mentioning the resource constraints or the environmental consequences of the various options. In this section we consider the issues of resource magnitudes and equity, any possible relationship of cooking to deforestation and global climate change, and indoor air pollution of alternative cooking systems.

Resources. Petroleum and coal are traded internationally, while local or regional self-sufficiency is usually sought for biomass resources. With the possible exception of ethanol, biofuels have low energy density and fuel transport can add substantially to the final cost (and reduce the net energy delivered by the fuel). Whether a specific biofuel can satisfy cooking fuel demand depends on the magnitude of the fuel resource and on competing needs. For biogas, demand for pumping water, generating electricity, and lighting gas lamps may compete with cooking. Sugarcane bagasse may be burnt as a fuel, gasified for generating electricity, or used as fiber for making paper.

Table 9.13 illustrates India's biogas resource constraint. India's cattle (including buffaloes) could provide about half of India's cooking energy needs during the peak season and a quarter of the needs in the lean season. These figures presuppose collection from all cattle, for example in community biogas plants. The potential is substantially lower for individual biogas plants, since not all households have sufficient cattle to meet their cooking needs, and excess gas produced by individual plants may not be available to other households. Thus, even in these favorable circumstances, biogas can only be a supplementary cooking fuel, and significant expenses for distributing the gas would be even harder to justify than indicated by the comparison in Table 9.11. Clearly, there are many "solutions" to this shortage, such as a dairy development program to improve the cattle stock (and dung yield), using other biomass to supplement dung in the digester, or even mixing wood-based producer gas to increase the energy content of the gas as necessary (Reddy, nd.). All of these alternatives require the additional commitment of land and resources for fodder or fuel purposes, which needs to be considered.

Table 9.14 compares land requirements for cooking with alternative wood-based systems (using fuelwood directly or as charcoal or producer gas) and with ethanol. We assume a typical yield of a tropical fuelwood plantation to estimate land requirements (in per six-person household) to provide cooking energy. While the absolute value of land requirements clearly depends on the yield, the relative land requirements for the wood-based systems are independent of yield. A producer gas-based system is the least land-intensive (170 m2 per household) while using a traditional stove with charcoal made in an inefficient kiln is the most intensive, requiring almost four times as much land. The best wood stove requires 40 per cent more and the best charcoal technology requires about 75 per cent more than the producer gas case.

Cooking with ethanol is the most land-intensive of all, although if the bagasse produces exportable electricity or fiber for paper and meets sugar mill and distillery energy needs, only a part of the land use should be charged to ethanol fuel production and the apparent land intensity would decrease. There are no additional land requirements for cooking with dung or biogas, since the animals primarily serve other purposes (food and traction). For the biogas case, the fertilizer value of the digester output is at least as much as the dung input so land requirement may in fact be considered to be negative.

Table 9.13
Biogas Resources for Meeting India's Cooking Fuel Needs
(1991 Populations: livestock, including buffaloes, 259 x 106; human 844 x 106)

	Peak season (5 months/yr)	**Lean season (7 months/yr)**
Dung yield/animal/day[a]	7.35 kg	3.52 kg
Dung production, all-India/day	1.90×10^9 kg	0.911×10^9 kg
National biogas potential/day[b]	1.48×10^{15} J	0.711×10^{15} J
Population whose cooking needs can be met	443×10^6	212×10^6
Part of total population	(52%)	(25%)

a At 35 liters per kg of fresh dung, and 22.5 MJ/M3 of biogas.
b Assuming a consumption of 7.33 GJ/year per household of 6, as in Ungra (See Table 9.11).
Source: Ravindranath and Chanakya, 1986.

Table 9.14
Primary Wood Requirements Using Fuelwood-Based Cooking and Land Requirements for Some Biofuel Alternatives

Fuel Stove	**End-Use Energy[a] (GJ/yr)**	**Primary Wood Energy[c] (GJ/yr)**	**Land Requirements (m^2/household)**
Fuelwood			
3-stone	20.27	20	330
Traditional	25.12	25	420
ASTRA ole	13.64	14	240
Swosthee MS4	17.15	17	290
Charcoal			
Traditional	14.49	40[d]	670
Traditional	14.49	24[e]	404
Improved *jiko*	11.00	31[d]	520
Improved *jiko*	11.00	18[e]	300
Producer Gas	7.33	10[f]	170
Ethanol	6.78	—	837[g]

a Numbers in this column are taken from Table 9.11.
b This column rounded to integers to facilitate comparison.
c Wood yield rate is assumed to be 29.7 dry tons per hectare per year, typical of plantations in tropical countries. (Lashof and Tirpak, 1989).
d Using a low-efficiency charcoal making process, such as the Tonga, with an energy conversion efficiency of 36 per cent (see Table 9.2).
e Using a brick beehive kiln BB1 with an efficiency of 61 per cent (see Table 9.2).
f Gasification efficiency of 70 per cent as assumed in the **Producer Gas** section.
g Brazilian distilleries working directly from cane yield 3,811 liters of 95 per cent ethanol per hectare per year (Ogden, et. al., 1990), for an energy yield of 81 GJ per hectare per year. There will be significant energy output in the form of bagasse as well as energy input in sugar refining and distillation that could lead to a surplus (Bernardo and Kilayko, 1990) but are not considered here.

Equity. Equity in access to biomass resources is a complex issue. Here, we present a few examples of cases where development policy needs to address equity issues.

In general, policymakers should make sure that the economic position of the poorest segments of society is not adversely affected by a development program. A program of subsidizing energy-efficient stoves in rural areas clearly is good from an equity perspective. Subsidies for individual biogas plants, on the other hand, generally go to the relatively well-off rural households which have the requisite number of cattle and the resources to make the rest of the investment. Whether or not this makes the poor worse off than before has been extensively debated, at least in India. On the one hand, people who originally gathered dung free as fuel may be denied it once biogas plants create a market value for dung. On the other hand, the significant environmental benefits of not burning dung may outweigh equity problems. Biogas plants may reduce the pressure on other cooking fuels (fuelwood and crop residues), making them more available to the poor. Evidence appears to favor the pro-biogas plant arguments (V. Joshi, TERI, pers. com., 1990). By converting more dung and making gas more widely available, community biogas plants should have less equity problems and greater environmental benefits than individual-household plants.

The end uses that are to be met by biofuels raise larger equity issues. For example, in Brazil, where alcohol is produced for use in automobiles, land could have been dedicated to growing food. Rosillo-Calle and Hall (1987) observed, however, that Brazilian food shortages and price increases resulted more from agricultural policies and economic problems than the fuel alcohol program.

Where a local biomass resource is used for cooking, generating electricity, etc., the equity problems are less, since the alternative centrally generated electricity and imported diesel incur higher social costs (Reddy et al., 1991; Bhatia, 1990).

Woodfuels and deforestation. Surveys show that each rural person uses about 1 kg of air-dried fuelwood equivalent per day (e.g., Smith, 1987). As about half of the world's population cook with biofuels, the total consumption is about 2.5 billion kg per day—almost a billion tons a year. Comparing the fuelwood demand with potential supply per country or for the whole world led to the identification of fuelwood as the "other energy crisis" and to the conclusion that wood gathering and land clearing were the "two principal causes of deforestation" (Eckholm 1975 and 1976).

Localized studies of patterns of biomass usage have led to different conclusions. In the Ungra area, rural households used mainly twigs, branches and roots for cooking—trees were not felled to meet local fuel needs (Ravindranath et al., 1991).

Using data from a rural energy survey (Natarajan, 1985), Leach (1989) showed that about 84 per cent of the total biofuel consumption in India "was in the form of crop residues, animal wastes or small branches and twigs, most of which ... has fallen to the ground" and do not contribute to deforestation. He concluded that "probably no more than five to seven per cent of total biofuel consumption involves permanent tree loss." Leach's conclusions have been borne out in other countries, and the Energy Research Group of Canada's International Research Centre concluded that "the use of firewood does not necessarily or frequently lead to deforestation" (ERG, 1986).

Conversely, fuelwood consumption in cities is derived almost entirely from whole trees and is likely to be a significant agent of deforestation (Reddy and Reddy, 1983). Demand for timber and other forest products is an important factor in the degradation of forests. Reddy (1987) distinguished between "degradation" and "loss" of forests. Degradation occurs from exploitation beyond a forest's capacity to regenerate, while loss occurs when forests are converted to other land uses. The two processes are different and need to be considered separately in formulating a policy to control deforestation.

While rural fuelwood users are seldom responsible for deforestation, they are often its victims. Wherever deforestation reduces available fuelwood, rural users have to adapt to the consequences. Many households in China, India, and elsewhere cook with crop residues or dung—fuels generally considered inferior to fuelwood. Burnt dung can not be used as fertilizer. Thus, improved cooking systems could alleviate fuel and fertilizer shortages.

Where cooking fuels contribute significantly to deforestation (e.g., urban and institutional cooking), improved cookstove programs are merited. As urban wood fuels are invariably purchased rather than gathered, it should be far easier to promote improved stove programs. Yet, some improved stoves programs (such as those in India) remain focused on rural areas. Because stoves can be built much faster than biomass can be "produced" through afforestation, improved stoves can delay deforestation and "buy time" for other means to halt or slow down deforestation (Meier and Munashighe, 1987). At the same time, biomass production needs to be increased to meet future needs.

Where urban use of fuelwood and other wood products is an important cause of deforestation, the nonrenewable felling of trees on public lands may be internalized by charging a "stumpage" fee that is verified at urban entry points. Openshaw and Feinstein proposed a method for determining economically-rational stumpage fees, a process that is being adapted successfully in the Sahelian countries (W. Floor, World Bank, pers. com., 1990). The stumpage fees do not affect rural gatherers, who are generally burning twigs and branches, rather than felling trees.

Indoor air pollution and health. The environmental consequences of cooking system choice range from air pollution in the kitchen to outdoor air pollution, acid rain, and global climate change. The cookstoves' contribution to outdoor air pollution in urban areas of LDCs and to acid rain has not yet been quantified. This section considers indoor air pollution; possible relationships between cooking technologies and climate change are considered next.

Air pollutants from cooking fuels include particulates, carbon monoxide, nitrogen oxides, sulfur oxides, and a variety of hydrocarbons. The quantity of each pollutant released by a cooking fuel is its emission factor (grams of pollutants per kg of fuel burnt), and depends on fuel type, and cookstove design and operation. The emission results in a concentration (mg/m3) of the pollutant in the air that depends on the kitchen size, ventilation, and the emission factor. A person's exposure to the pollutant depends on the time spent in the kitchen during cooking. The dose depends on pollutant characteristics. The health effects depend on the toxicity and dose of the pollutant. This sequence of relationships is graphically represented in Figure 9.10.

Although respiratory diseases are the main cause of death in developing countries (WHO, 1984) there has been little systematic study of the health effects of air pollution generated by cooking stoves. Smith (1987) conducted an excellent review of studies on air pollutation and health implications of biofuel cookstoves.

Concentrations of total suspended particulates, benzopyrene, and other pollutants in rural kitchens of developing countries often exceed by large margins the ambient air quality standards of these pollutants in all countries. In the absence of epidemiological studies, and contrary to the usual practice, health effects must be estimated by extrapolation from low to high doses.

A careful series of epidemiological studies conducted in Nepal demonstrated statistically significant relationships between domestic smoke exposures and chronic bronchitis, chronic cor pulmonale, and acute respiratory infection (Pandey, 1984; Pandey et al., 1985 and 1989). Studies in other countries have not led to consistent interpretations, perhaps because of "different definitions of the biological endpoints, such as chronic bronchitis" (Smith, 1987).

While few epidemiological studies of the health effects of rural cooking in LDCs exist, no studies have been undertaken to document the health impact of improved stoves programs.

TERI conducted a series of laboratory tests measuring emission factors and emissions per task of carbon monoxide (CO) and total

Figure 9.10

Categorization and Flowchart of Separate Topics Involved in Investigating the Extent and Impact of Air Pollution Exposures from Combinations of Biofuels in Developing Countries

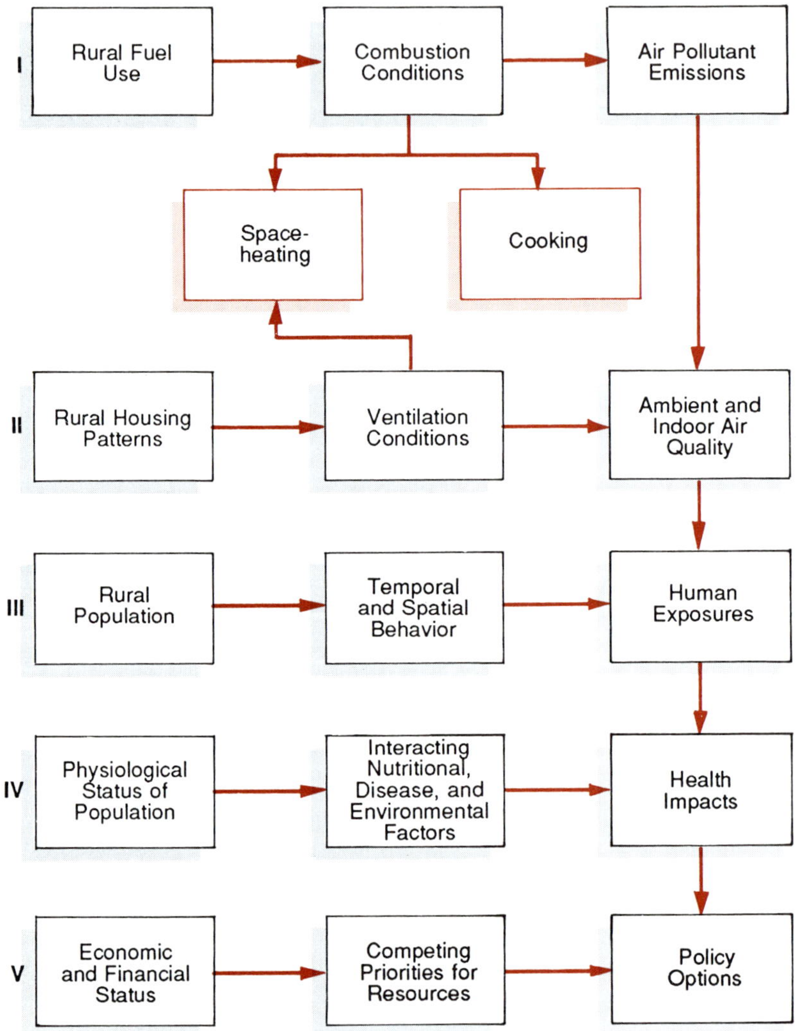

suspended particles (TSP) for different stove and fuel combinations (Ahuja et al., 1987; Joshi, Venkataram and Ahuja, 1985 and 1987; TERI, 1987). TERI (1987) "confirmed that emission factors for biomass fuels increase with efficiency, but in half the stove-fuel combinations, increase in emissions was offset by the increase in efficiency."

Table 9.15 summarizes laboratory measurements of emissions per task of CO and TSP in seven biomass stoves and two kerosene stoves. Crop residues and dung generate far more pollutants than fuelwood. Surprisingly, the kerosene stoves produced almost as much CO and TSP as the cleanest fuelwood stoves. Similar tests were carried out on two different two-pot mud and cement stoves with chimneys. Venting the chimney outside the room removed 30-88 per cent of the CO and 62-75 per cent of the TSP.

Table 9.15
Emissions per Task of Carbon Monoxide and Total Suspended Particulates (TSP) from Seven Metal Biomass and Two Kerosene Stoves

Stove	CO (g/task)				TSP (g/task)			
	Fuel-wood	Mustard Stalk	Dung	Kero-sene	Fuel-wood	Mustard Stalk	Dung	Kero-sene
Metal biomass stoves								
a	5	11	14	–	0.6	1.2	2.2	–
b	5	16	–	–	0.3	1.7	–	–
c	6	13	11	–	0.7	1.3	1.4	–
d	9	–	–	–	0.4	–	–	–
e	9	–	–	–	0.6	–	–	–
f	9	20	–	–	0.5	1.4	–	–
g	11	21	–	–	0.5	1.3	–	–
Kerosene stoves								
a	–	–	–	4	–	–	–	0.2
b	–	–	–	4	–	–	–	0.2

Note: The cooking task is simulated by raising temperature of 3.5 liters of water to 60°C.
Source: TERI, 1987, Tables 4.D.11 and 4.D.12.

Using a portable monitor worn by the cooks, Ramakrishna (1988) measured CO concentrations and TSP exposures in sample houses using traditional open stoves and improved stoves with chimneys. While the pollutant levels were somewhat lower with the new stoves (Table 9.16), the difference was statistically significant ($p < 0.05$) in only two of the three villages. The relatively small reduction in stoves with chimneys confirms that these stoves are not built or maintained as intended (Smith, 1987). Substituting pots in pot-specific stoves and leaving unused pot holes open can also increase smoke release into the kitchen.

Pending further epidemiological studies of the health effects of traditional and improved stoves, TERI (1987) proposed that any new stove should give lower pollutant emissions per task than existing models, and that efficiency should improve. Clearly, laboratory measurements of emissions per task need to be conducted on other stove/fuel combinations. Because of the differences between laboratory and actual kitchen conditions, pollution reduction goals need to be verified through field measurements.

Biofuels use and climate change. Of the two important climate change concerns—global warming and the depletion of the ozone layer—only the first is affected directly by cooking technology choice. Carbon dioxide (CO_2) and methane, both emitted by some cooking processes, contribute to global warming (Mintzer, 1978).

If biomass is used renewably there is no net CO_2 production because an equivalent amount is absorbed in tree growing. Fossil

Table 9.16
Carbon Monoxide and Total Suspended Particulates in Village Kitchens Using Traditional Stoves and Improved Stoves with Chimneys

Stove Type	Sample Size (no. of stoves)	Burn Rate (kg/h)	Mean CO Concentrate (mg/m^3)	Mean TSP Exposure (mg/m^3)
Garhibazidpur Village				
Traditional	34	1.4	7.0	3.3
Sahyog	22	1.4	6.0	2.9
Keelara and Pallerayanahalli Villages				
Traditional	24	1.9	19.0	3.2
Traditional with hood	16	2.0	5.7	1.7
ASTRA ole	20	2.1	8.9	2.5

Source: Ramakrishna, 1988.

fuels produce amounts of CO_2 that depend on the carbon content of the fuel. In Table 9.17, we compare CO_2 emissions for an Ungra six-person household, using a variety of stove and fuel combinations. For fuelwood, charcoal, and producer gas, emissions range from zero (corresponding to renewable use of wood) to an upper limit (if there is no replanting to replace burnt wood). As biofuel use has not been a significant source of deforestation, the contribution of biofuel use to CO_2 emissions must be relatively minor. In heavily deforested parts of Asia, crop residues and dung are cooking fuels. Crops (including sugarcane) are replanted (renewable) and therefore produce no net CO_2; similarly, with a stable cattle population, steadily burning dung or biogas produces no net CO_2. The comparison in Table 9.17 includes natural gas and coal, which represent the lower and upper CO_2 emission factors among commonly used nonbiomass cooking fuels.

Methane emissions also make a significant contribution to global warming. Table 9.18 shows global and Indian emissions of methane, by source. The digestive tracts of herbivorous animals are one of the largest sources of methane. Part of their methane can be converted to carbon dioxide if burnt as biogas. The conversion will reduce the

Table 9.17
Carbon Dioxide Emissions in Cooking with Selected Cooking Systems

Fuel	Stove	Annual Fuel Use[a] (GJ/HH)	CO_2 Emission Factor[b] (kg CO_2/GJ)	Total Emission of CO_2 (kg/household per year)
Wood	Traditional	25.12	0-85	0-2135
	ASTRA	13.64	0-85	0-1159
Charcoal	Traditional[c]	40.25[d]	0-85	0-3421
	Improved	18.03	0-85	0-3421
	Producer gas	10.47[e]	0-85	0- 890
Dung, biogas, crop residues, and alcohol: no net production of CO_2				
Kerosene	Nutan	6.35	77	489
LPG	Superflame	5.40	59	319
Natural gas		(5.40)[f]	51	275
Coal		(14.49)[g]	95	1,377

a Fuel consumption data taken from Table 9.12.
b CO_2 emission factors taken from DeCicco et al., 1990.
c Charcoal made in a 36 per cent efficient kiln and burnt in a traditional stove.
d For charcoal and producer gas, the primary fuelwood consumption is shown here and used to estimate CO_2 emissions.
e Charcoal made in a 61 per cent efficient kin, and used in an improved jiko.
f Energy consumption with natural gas is assumed to be the same as with an LPG stove.
g Energy consumption with coal is assumed to be the same as with charcoal in a traditional stove.

Table 9.18
Methane Emissions in India and the World
(million tons/year)

Source	Global	India
Animals	76	10.3-10.4
Rice	114	32.8-33.8
Biomass burning	65	4.3
Coal mines	35	1.3
Natural gas	47	0.1
Landfills	41	2.5-2.9
Total	**378**	**51.3-52.8**

Source: Pachauri and Suri, 1990.

impact on global warming since a molecule of methane is about 25 times as effective in absorbing infrared radiation as a molecule of CO_2. It is difficult to estimate how much methane is burnt off since the methane-converting bacteria produce additional methane that would not be produced if the dung were left alone. (India's biogas potential, estimated in Table 9.13, is equivalent to 7.2 million tons of methane, assuming biogas to be 55 per cent methane by volume. Most of the remainder is CO_2, which leaves open the possibility of sequestering it, particularly in large digesters.)

The methane emitted from biomass burning contains a significant quantity of energy, and contributes to global warming. Thus, the design of biofuel burners deserves further research. Methane is also produced in landfills. The estimated 2.7 million tons (Table 9.18) released in Indian landfills contain 135 x 1015 J of energy, enough to satisfy the cooking energy needs of 150 million people. Methane from worldwide landfills could supply the cooking energy needs of 2.3 billion people. Coincidentally, this is roughly the number of people who currently depend on biofuels for cooking. Capturing and burning landfill gases would reduce the atmosphere's methane burden while providing a significant amount of energy.

Summary

The choice of a cooking system is complex. The system must satisfy multiple criteria for the users (ease of turning on and off, control of heat output, speed, and smokelessness). The stove and associated equipment should be economical from the user's perspective, often implying low purchase cost. The fuel supply should be reliable, and fuel cost affordable. At the national level, foreign exchange implications, renewability of energy source, environmental soundness, and meeting overall development objectives are important issues.

In this chapter, we reviewed user fuel preferences, energy consumption, economics from a societal perspective, resource constraints, and environmental implications of biofuels (which also reduce energy use compared to solid fuels). With kerosene at the equivalent of $30 per barrel of petroleum, biofuels are the least cost alternative from a societal perspective. Because of tax and subsidy policies and wood gathering from common lands, the individual economic perspective may be very different. Generally, policies should be adjusted in order to attain the societal optimum. The following policy instruments are proposed:

- *Marginal cost pricing.* Subsidies that price fuels and electricity below their replacement cost should be gradually removed. Electricity prices should reach long run marginal costs; where applicable, petroleum fuels should be priced at international levels with an appropriate import premium and stumpage fees charged for wood at urban entry points.

- *Financing capital investments at interest rates equal to social discount rates.* These interest rates would be much less than the implicit discount rates of low-income households. No subsidy is involved and upper-income users may wish to purchase the equipment outright or with alternative financing.

- *Subsidizing energy-efficient stoves.* Rural raw biomass users often depend on gathered wood, dung, or crop residue and thus have little economic incentive to invest in an energy-efficient cookstove. Moreover, they may not have the necessary capital. If improved biofuel stoves or other alternative cooking systems would improve village life by reducing time spent gathering fuel and/or cooking, making more dung available as fertilizer, reducing indoor air pollution, etc., then some form of subsidy will be necessary, at least for lower-income households. An "improved" stove deteriorates physically and its fuel use increases with time so that it will need to be replaced after one or two years. Subsidy programs should take this into account if the impact is to be permanent.

Many technical opportunities exist for reducing cooking energy use by improving the efficiency of biofuel production and of cookstoves. Fuel is often the largest component of total cooking costs, so that substantial effort is justified to develop energy-efficient stoves for all fuels. Fuel economy is best compared through controlled cooking tests, while the overall impact of an alternative cooking system needs to be evaluated using field measurements of fuel use.

Land and other resource requirements of different biomass-based alternatives should be considered before promoting a new cooking system. When a particular resource cannot meet all cooking energy needs and requires significant infrastructure (e.g., piping), the economics of the alternative can be greatly altered.

Indoor air pollution from cooking is a significant health hazard, particularly in LDCs. An alternative cooking system should not generate more air pollutants per cooking task than the system it is intended to replace. The pollutant generation is best measured in conjunction with controlled cooking tests wherever possible.

Several biofuel alternatives (dung, crop and fuelwood residue, biogas, and alcohol) generate no additional CO_2. If used in a renewable manner, fuelwood and derived fuels do not produce any net CO_2 either. In India, only 5-7 per cent of total biofuel consumption may represent "permanent" tree loss. Using biogas and methane from landfills for cooking or other purposes reduces methane's contribution to global warming.

We have presented a technical and economic evaluation of alternative cooking systems in this chapter. Social and cultural factors play an important role, particularly in rural areas outside the monetary economy. Among indigenous people, food preparation is generally of ritual significance that lies beyond the quantification illustrated here. If we are to avoid a permanent loss of the planet's rich cultural diversity, steps must be taken to utilize available energy resources on a sustainable basis, while taking into account the beliefs and customs of the world's peoples.

10

Energy Efficiency Barriers, Intervention Strategies and Current Programs

The existence of large and cost-effective opportunities for electricity conservation in India raises the question: Why haven't these cost savings been taken advantage of already?

Barriers to the adoption of energy-efficiency measures need to be overcome by appropriate intervention strategies. In this chapter we consider the barriers and then discuss a variety of intervention strategies for overcoming them.

Some barriers are common to saving any form of energy in all countries. These common barriers are (see e.g., Geller, 1989a):

- lack of information or awareness among consumers, especially about
- recently available or rapidly evolving technologies;
- uncertainty regarding savings and cost-effectiveness, as well as future energy prices;
- lack of capital and resistance to buying equipment with a greater purchase cost;
- requirements for rapid payback and high rate of return of building owners, occupants, businesses and industries; and
- separation of responsibilities for making capital investments and paying operating costs. For instance, some energy-consuming equipment (heating, air conditioning, water heating, refrigeration, etc.) in US apartment buildings are provided by the landlords while the tenants often pay for the energy used by this equipment. Many new single family houses are sold equipped with a variety of appliances. The builder does not pay the energy bills and has no incentive to purchase anything but the equipment with the lowest purchase cost.

These obstacles to energy efficiency are present in developing countries but other barriers can be even more important.

Energy planners are often unaware of the large potential for energy savings in developing countries. Referring to the relatively low level of commercial energy use in many developing countries, analysts believe that the only possibility for development lies in the direction of energy supply expansion. Even some analysts from industrialized countries believe that the potential for energy conservation in developing countries is small. For instance, in an analysis of the energy/capital crisis facing developing countries, McKillop (1986) stated:

> *... These (structural) factors, together with the enormous debt burden, must serve to reduce the frankly naive projections of some observers for the potential role of energy conservation and substitution programs in LDCs where 30 per cent cuts in oil demand are lightly presented as both feasible and cheap... The mirage of energy conservation, while being an apparently attractive concept for policy and related studies, has little immediate role to play in the energy economics of LDCs...*

More specifically, World Bank energy economist Munasinghe (1983) had noted that:

> *(The) reduction in relative value of conservation will always occur with increases in the discount rate, because increases in initial investment are traded off against the future savings... Energy users who confront high opportunity costs of capital, for example those in many LDCs, will find costly capital-intensive energy conservation measures relatively less attractive than users who have access to low-cost sources of capital...*

In response to Munasinghe's valid concern, Williams (1988) comments:

> *The difficulty consumers have in obtaining capital in a developing country should be distinguished from the problem of the country's overall limited supply of capital. Under a wide range of circumstances, the extra capital requirements for improved end-use technologies will be more than offset by capital savings for energy supplies that would not be needed.*
>
> *Not only are investments in energy efficiency relevant for developing countries, but also they are often even more relevant than for industrial countries. This perhaps counter-*

intuitive result is illustrated in Figure 10.1, which compares, for the Brazilian situation, the cost of saving a kW with investments in more energy-efficient end-user devices (more efficient refrigerators, compact fluorescent light bulbs as alternatives to incandescents, and variable speed drives retrofitted onto alternating current induction motors) with the cost of the extra hydro-electric supply expansion that would be required if less-efficient conventional end-use technologies were used instead. The cost shown in each case, as a function of the discount rate, is the discounted present value of all required investments over a 50-year life cycle. In the light bulb case, as an example, for a 10 per cent discount rate the supply expansion cost per kW is 2.4 to 3.2 times the cost of saving a kW by investing in more efficient bulbs. What is perhaps more interesting, is that in this case and the other cases as well, the net benefit of the energy-efficient option increases rapidly with the discount rate.

Thus, from a societal perspective, investments in energy efficiency improvement look better, the higher the discount rate. This analysis leads to a rather surprising result: contrary to the concern expressed by Munasinghe, investments in energy efficiency improvement will often make even more sense in capital-short, developing countries than in industrial countries (emphasis added).

An analysis of the electricity savings opportunities for Brazil ... for just six end-use activities (domestic, commercial, and street lighting, domestic refrigerators, more efficient motors, and variable speed drives for motors) indicates that investments in efficient technologies in these areas could reduce electricity demand growth, 1985-2000, from the 6.4 per cent per year rate forecast by the government to 4.8 per cent per year. As a result the need for new power plants would be reduced in this period by 19 GW, saving some $38 billion in capital requirements for new electrical capacity but requiring an investment of some $8 billion in more energy-efficient end-use technologies (Geller et al., 1988; emphasis added).

(Note Reddy's analysis for Karnataka [summarized in Chapter 7, see Figure 7.3]. The least-cost electricity plan—consisting of four conservation measures and three decentralized, renewable generation options—required capital investments of about $0.6 billion a year as opposed to $1.6 billion for the conventional plan which considers

Figure 10.1

Cost of Electricity Generated by New Hydroelectric Plant vs Cost of Electricity Saved Through Energy Efficient Options

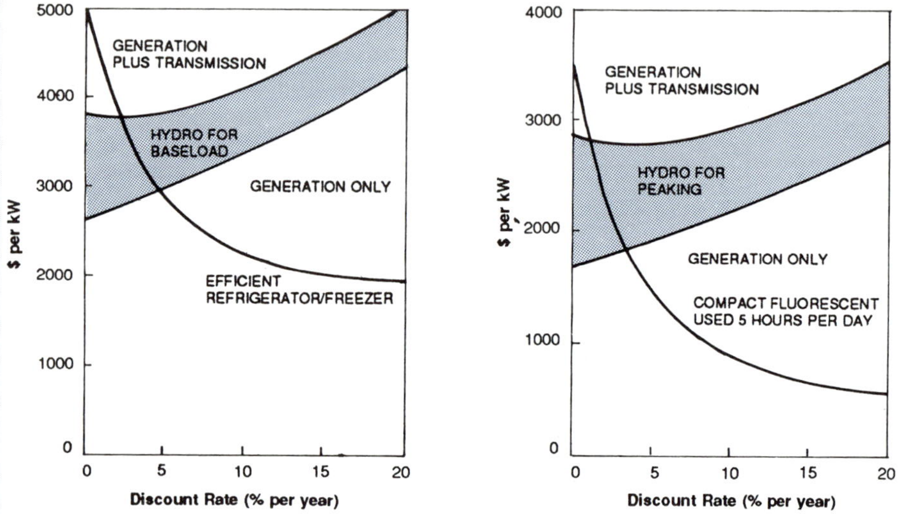

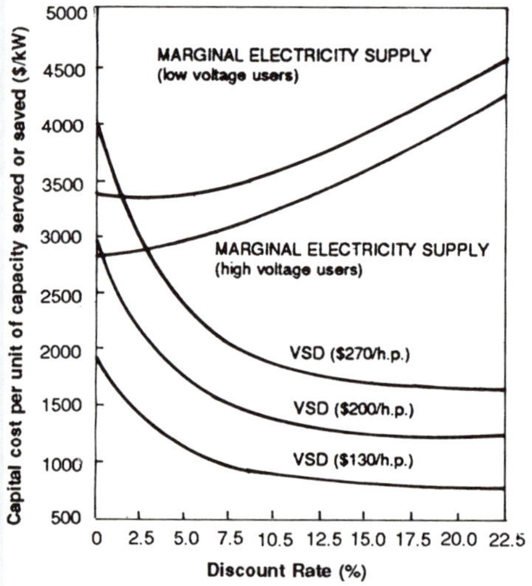

Discounted present value of the cost of electricity produced with new hydroelectric power supplies vs. the cost of electricity saved via the purchase of more energy-efficient refrigerators as alternatives to conventional refrigerators (left), the installation of energy-efficient compact fluorescent light bulbs (middle) as alternatives to incandescents (see caption to Figure 9), and the installation of variable speed drives on AC induction motors (right), as a function of the discount rate.

only supply expansion using [centralized] coal, nuclear and large power plants.)

While the potential for significant energy savings is slowly being recognized by planners in developing countries, many established policies present obstacles much less common in industrialized countries. One such barrier stems from subsidies of energy prices. While subsidies are usually larger for petroleum products and are generally larger in oil-exporting developing countries (Figure 10.2), significant subsidies are often found in oil-importing developing countries as well, particularly in the case of electricity.

Kosmo (1987) defines an "economic" subsidy for electricity as:

> *... the long-run marginal costs of providing service minus the domestic price, where marginal costs represent the costs of providing one additional unit of energy. Accordingly, if the long-run marginal costs of providing electricity are $0.10/kWh while the price is only $0.07/kWh, then the economic subsidy is $0.03/kWh, even though the utility might not need any financial subsidies as long as average costs are less than $0.07/kWh.*

Residential and agricultural tariffs in India carry large economic as well as financial subsidies. Table 10.1 shows average electricity prices for agriculture, residences, and industry, as well as the average cost of electricity generation and supply. The marginal cost of electricity is around $0.09/kWh. One consequence of heavily subsidized electricity for agriculture has been the large number of electric pump sets in India compared to other countries. More importantly, well-documented and significant energy conservation measures are not cost-effective for pump users, though they have very low cost of saved energy of about $0.01/kWh.

Since electricity cannot be inexpensively stored in large quantities, demand and supply must always be matched. Hourly or seasonal variations in electricity use result in demand peaks that are met by more expensive supply options compared to off-peak demand. With few exceptions, developing country electricity users do not pay a time-differentiated electric price. Thus, measures to shift demand to off-peak periods (load management) are not cost-effective to the user, although the shift may reduce the overall cost of supplying electricity, reduce the possibility of shortages, or both.

The lack of available energy-efficient products limits the possibilities of energy conservation. Many products available internationally are not yet manufactured in LDCs. High import tariffs or non-fiscal barriers may stop energy-efficient products from entering a domestic

Figure 10.2
Changes in Gross National Product and Energy Use Japan, 1965-87

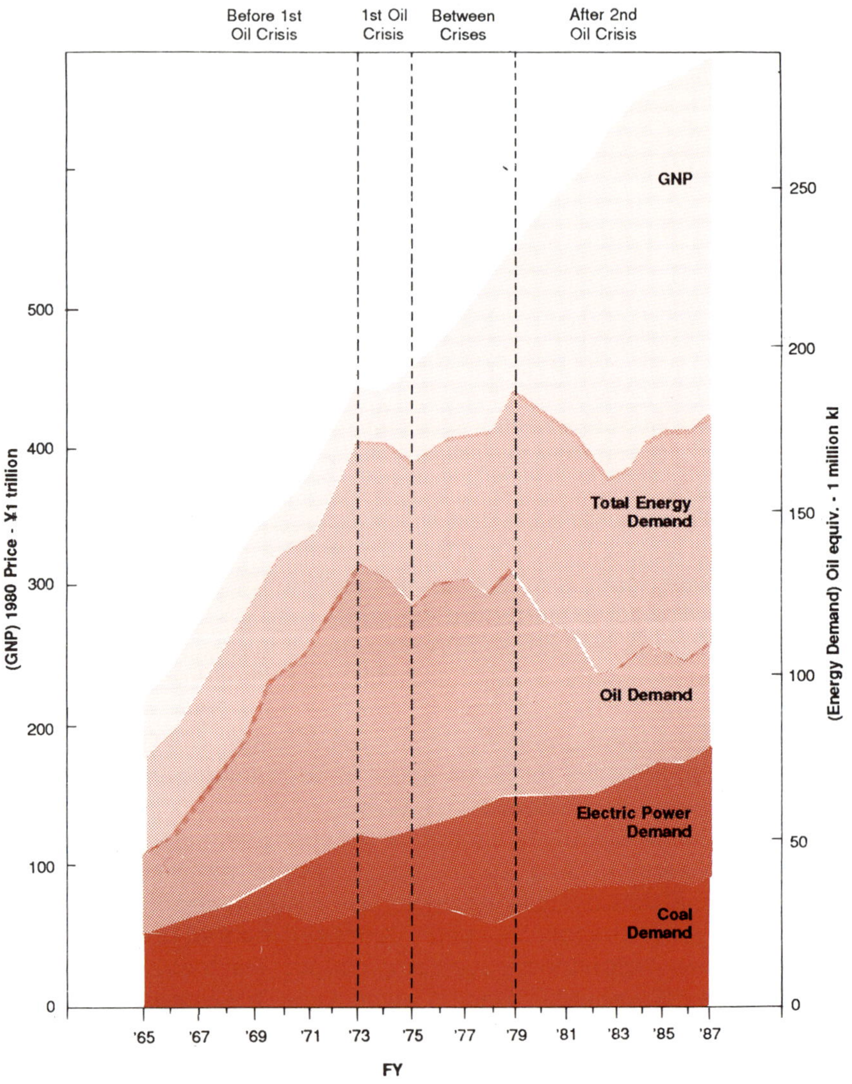

Table 10.1
Average Indian Electricity Prices for Agriculture, Residences and Industry, and Average Costs of Generation and Supply (1988-89)

End Use	Average Price (Rs/kWh)	Equivalent (1990 $/kWh)
Agriculture	0.16	0.009
Residences	0.54	0.032
Industry	0.98	0.058
Average cost of generation and supply	**0.91**	**0.054**

Source: *Current Energy Scene in India,* 1990.

market. Such market protection gives little incentive to local manufacturers to innovate on energy efficiency (e.g., refrigerators made in India). Even when certain energy-efficient products are manufactured nationally, the small market volume may result in exorbitant prices (e.g., reflective luminaires and energy-efficient motors in India). Sometimes, the locally manufactured product may be inferior (e.g., electronic lamp ballasts and fluorescent lamps). The cost-plus pricing policy of some (many?) public sector industries also provides little economic incentive to save energy.

A least-cost electricity plan will involve major shifts in capital investments from power plant construction to end-use efficiency improvement. Power plant suppliers and installers represent powerful vested interests who will try to resist this shift in capital investments. Since engineering and management skills required for installing conservation measures and cogeneration projects on a large scale are similar to those required for designing and building power plants, it would be far better if the contractors join, rather than resist, the new orientation to providing energy services.

Barriers to energy conservation common to all countries, and those specific to individual nations, must be overcome to attain the full potential of electricity efficiency, which brings with it significant economic benefits and greatly reduces the environmental consequences of electricity generation. Many industrialized countries have significantly increased energy efficiency, separating economic growth from energy demand. Japan may present the most remarkable example (see Figure 10.2). Based on the experience of industrialized and some developing countries, intervention strategies to overcome the barriers to increased efficiency are presented below.

Improved Information

An essential step in improving energy efficiency is the improved flow of information. Available information should reach potential audiences and research and development should endeavor to fill important gaps and improve the quality of the information. We discuss below the elements of information needed, who could use this information, and how it may be made more accessible.

Data on electricity end uses. Before planning or executing a major electricity program, we need to know where the energy uses are concentrated. Traditionally, energy sales data are disaggregated by sector of users, e.g., industry, agriculture, residences, etc. For electricity, there is another breakdown by tariff category. While this is useful, it is insufficient to tell us where to find the energy savings are to be found.

To begin to estimate the potential for energy savings we need to know how much energy is consumed for each end use, what devices are used, and what is their energy efficiency. Earlier in this report, we tried to disaggregate electricity use in India by end uses with some information on the type and efficiency of the end-use devices. The lack of data meant that our analysis is crude. The next step would be to improve the data base.

Surveys need to be conducted in each sector to determine the stock of electricity-using equipment and establish its energy consumption characteristics, preferably through direct measurements. In some cases, surveys have been carried out, e.g., energy use in households in a few cities and for electric motors in major industries, but in general surveys fall far short of providing representative samples of the entire country. The surveys should be done regionally, and later aggregated to a national level to permit regional matching of energy supply and demand. These surveys need to be repeated every few years to update information.

Compilations of energy data are published annually by TERI *(Energy Data Directory and Yearbook),* by the Centre for Monitoring the Indian Economy *(Current Energy Scene in India),* and by a variety of government agencies. These organizations could include end-use data breakdown in the future, if such data were available. Some of the preliminary end-use disaggregation presented in this report has been obtained from data published by these organizations.

Data on costs of electricity generation alternatives. To calculate the cost-effectiveness of electricity conservation and load management opportunities, and to develop a least-cost plan for providing energy services, we need to know the capital, operating, and maintenance costs of alternative generation technologies. Chapters 7 and 8 illustrate how this type of information was used to develop a least-cost electricity plan for Karnataka.

Opportunities and magnitude of electricity savings. Energy-efficient alternatives to present end-use devices indicate the possibilities for energy savings. An economic calculation (Chapter 2) can be used to determine whether an energy-efficient alternative is cost-effective. Then, considering all such energy conservation opportunities and the total energy currently being used for specific end uses, one can estimate the potential for energy savings by end use. Once the breakdown is available, energy conservation programs may be directed towards end uses that represent large or particularly cost-effective energy savings. A preliminary estimate of conservation potential by end uses has been presented here. The estimate needs to be refined using better end-use data, but clearly points out some areas deserving immediate attention. These include irrigation pumping, electric motors in industry, residential appliances, and residential and commercial lighting. Programs may be initiated or furthered in these areas while the end-use data base is being upgraded.

Scorekeeping electricity conservation. As the extent of breakdowns in electric power plants affects the cost of electricity generation, the cost-effectiveness of energy conservation measures is ultimately determined by energy savings that are realized in actual use. Energy savings realized from past conservation efforts should be measured in a sample of projects and compared with expected savings in order to improve future energy-saving projections and modify conservation strategies, if necessary. Procedures for estimating electricity savings using electric billing and other readily-available data before and after implementation of an energy conservation program are referred to as scorekeeping (Dutt and Fels, 1989).

Testing and certification facilities for energy-efficient products. Potential users cannot begin to choose an energy-efficient device, however cost-effective, if they do not know the energy efficiency of alternative models available. While the energy efficiency of residential appliances, electric motors, etc. is routinely tested by manufacturers in industrialized countries, such testing is virtually nonexistent in developing countries. The Bureau of Indian Standards (BIS) has developed standards that specify minimum energy efficiency and other characteristics and test procedures for a wide range of products (BIS, 1990). While some manufacturers and government departments (e.g., the Central Power Research Institute and the Directorate General of Supply and Disposal) are equipped with testing facilities, testing appears to be sporadic and there is no published compilation of the energy performance of the available models of refrigerators, motors, pumps, etc. Often the testing is left to the manufacturers, and test results reported by BIS or other independent verifications of manufacturers are not published. Moreover, manufacturers not wanting to meet the standards (which enable them to put the Indian Standards Institution mark on their products) do not have to test their products at all. As of March 1988, only three certification mark licenses had been issued for refrigerators, ten for air conditioners, and seven for ceiling fans and regulators (Paramesvar, 1988, Annex 3). The BIS is interested in the whole set of norms that make up a given standard, not just the energy-efficiency aspects.

Clearly, what is needed is a properly-equipped independent facility that tests all models available for sale in the country. The testing could be initiated with the end-use devices judged to be most important (e.g., domestic refrigerators, air conditioners, lamps, electric motors, pumps), and later extended to other devices. The laboratory would certify the energy consumption of efficiency (under test conditions already specified by the BIS) of the various products tested. Clearly, different laboratories could specialize in different kinds of products. In a precedent, in India's Improved Cookstoves Program, biomass-burning stoves qualifying for government subsidy must be tested and certified in one of several government-approved laboratories. As noted in Chapter 4, the Indian Pump Manufacturers' Association plans to test pump performance within the framework of an Indian Hydraulic Research Centre. Test data could be compiled and made available to manufacturers and user groups, and could be the basis for establishing or upgrading minimum efficiency standards (discussed below).

Training of energy auditors and energy managers. End-use energy analysts can identify the overall potential and cost-effectiveness for different conservation measures, but implementing energy conservation projects will require analysis at the individual user level. These energy audits will need to be performed on thousands or tens of thousands of industries, office buildings, etc. Larger installations will require qualified energy managers on staff. Japan's success in energy efficiency has depended on highly-trained energy managers (see Fig. 10.2). At present relatively few companies are conducting energy audits in India and other developing countries. The Industrial Development Bank of India (IDBI), which currently provides subsidies for energy audits, lists 72 approved energy consultants (V. Bablani, IDBI, pers. com., 1990). By October 1990, IDBI had sanctioned only 150 energy audits. Although the National Productivity Council and other groups have conducted many more audits outside the IDBI scheme, the number of audits completed in relation to that needed is small. The Energy Management Centre (EMC), set up in 1989 with support from the European Economic Community (EEC), notes the shortage of energy managers and auditors, and has started a training program to help alleviate the shortage (EMC, 1989). Details for the EMC are given in Box 10.1. Apart from initial training, energy auditors need to be apprised of new technologies and conservation opportunities through ongoing training updates, newsletters, etc. The monthly magazine Energy Management published by the National Productivity Council and energy conservation case studies published by TERI are good examples of information dissemination activities.

Information to manufacturers about improving product efficiencies. As noted, manufacturers in developing countries do not produce many of the newer energy-efficient technologies which are developed and available internationally. Manufacturers need better access to what is available. The best way to channel this is through appropriate manufacturers' associations. The associations in India include:

IPMA	The Indian Pump Manufacturers Association
IEEMA	Indian Electrical and Electronic Manufacturers' Association
AIEMMA	All India Electric Motor Manufacturers Association
ELCOMA	Electric Lamp and Component Manufacturers Association
CEI	Confederation for the Engineering Industry (an umbrella organization)

In Chapter 4 we noted the IPMA initiative of improving the capabilities for designing and manufacturing efficient pumps. Support needs to be channelled to these and other efforts to improve the manufacturing facilities and provide the foreign exchange requirements for appropriate licensing agreements.

Information to users about costs and savings of alternatives. Energy users need to know something about the potential for energy savings, how to go about it, and benefits for them to initiate any action. Undirected announcements to "Save Energy" and "Save Petrol," abundant in the Indian media and on billboards, are unlikely to have any impact. Indeed, there is a danger that people will become anesthetized to calls for saving energy so that when more focused information is presented, it may be met with indifference ("There they go again, telling us to save energy"). Specific information, targeted to particular user groups, may be useful, provided other barriers are addressed as well.

One type of information particularly suitable for industries is a "technology menu" for conservation opportunities. The Technology

Box 10.1
Energy Management Centre

The Energy Management Centre (EMC) was established in 1989 under the Department of Power, Ministry of Energy, Government of India with the cooperation of the European Economic Community. Although there are many institutions active in energy management, there is fragmentation of effort, and lack of coordination in the various initiatives that are under way. EMC was established to provide an overall strategic vision, based on quantitative analysis, of what needs to be done to improve energy efficiency in the Indian economy.

"The Centre aims to develop innovative methodologies in energy management techniques, particularly the software aspects, to increase the capability of energy professionals. Along with improving the energy information system and data base, the Centre will train energy managers and experts in the country. It will organize the exchange of research findings on energy policy, management, planning techniques, statistical assessments, forecasting, etc. among institutes, organizations and industries within the country as well as between Indian and European professionals.

"The Centre intends to play a significant role in the evolution and implementation of programmes that would strengthen energy management in the country. It will promote innovative management practices and systems in the energy sector and organize exchange of experts between India and the EEC and between India and other countries in the Asia-Pacific region. In addition, the Centre will establish or assist in the establishment of sub-centres across the country to utilize local talent and implement its programmes."

(Source: *Energy Management News,* No. 1, Nov. 1989).

Menu concept, developed by Larson of Princeton University's Center for Energy and Environmental Studies, divides electricity consumption according to ten end-use categories. The menu for one such category, "movement of material" (motors, pumps, fans, ducts, tubes, solids conveyors, etc.) was developed for Sweden in 1989 (Larson and Nilsson, 1989). The menu includes a brief description of the technology alternatives of each end-use system including its components, energy savings potential, and approximate economics. For developing countries, additional information would be useful, particularly for smaller industries with little in-house energy management capability. Information about local availability, suppliers, energy audit consultants, financing sources, and incentives should be included.

For residential, commercial, and agricultural users, with fewer different end uses, brochures and television programs can help inform people about conservation opportunities. In December 1990, Indian television aired an excellent documentary on energy efficiency alternatives for lighting, including light sources, luminaires, and controls. Displays in trade and other fairs, which are well-attended and popular, can help disseminate specific energy conservation opportunities. Some of the information is well-suited to secondary school science curricula, although in India these have so far resisted all attempts to be meaningful.

Financial institutions (banks, development financing agencies) are likely to play an important part in disbursing the financing to implement a wide range of energy conservation projects and providing financing for purchases that have future energy implications. Seminars to keep bank officials informed about life-cycle costing of energy using durables and about energy conservation investments that are particularly cost-effective should help to overcome any resistance to giving larger loans (corresponding to a higher purchase cost device) for options that have lower life-cycle cost if energy costs of operation are included.

Economically Rational Energy Pricing and Fiscal Policies

To adopt energy efficiency, users need an economic incentive. As noted above, pricing policies, tax, and import policies often reduce or eliminate the economic advantage in saving energy. These policies should be reviewed in order to make sure that relative subsidies and taxes are conducive to minimizing the social costs of providing energy services. Some elements towards economically rational fiscal policies are discussed below.

Marginal cost pricing. Energy subsidies are often promoted as an instrument for economic growth and reduced inflation. On the basis of a statistical correlation (Fig. 10.3), Kosmo (1987) showed that

Figure 10.3

Real Growth Rates of Gross National Product and Retail/Border Price Ratios Annual Averages, 1973 to 1983**

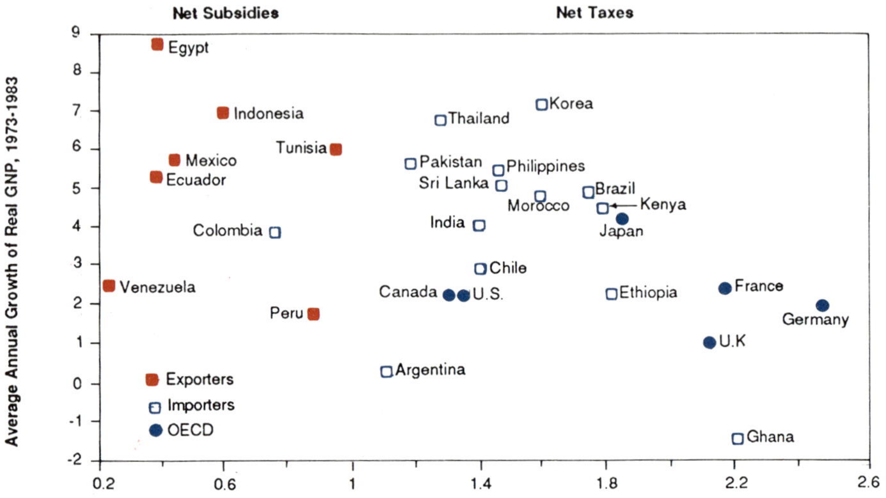

** Retail prices are a weighted average (by consumption shares) of the prices of the four primary petroleum products - gasoline, diesel oil, kerosene, and heavy fuel oil. Ideally, these retail prices would take into account the consumption of natural gas, electricity, and coal since in a few countries (e.g. coal in India and Korea) these fuels account for substantial shares of consumption. However, data limitations make this impossible. In any event, changes in petroleum prices are the best proxy for changes in retail energy prices - for the 22 OECD countries, from 1978-1984, the correlation coefficient between domestic petroleum prices and domestic energy prices is .89. In the developing countries, the correlation is likely to be higher since they are relatively more dependent on oil for commercial energy needs.

Fitted Regression Line	95% Confidence Interval for Slope of Line	Correlation Coefficient (x,y)
y = 5.82 - 1.01x [Exporters] (3.95)	-10.67 , 8.66	-0.11
y = 6.87 - 1.91x [Importers] (1.82)	-5.84 , 2.02	-o.29
y = 3.14 - 0.38x [OECD] (1.08)	-3.17 , 2.41	-0.17

y = Average Annual GNP Change
x = Average Retail/Border Price
* = Statistically Significant at 5% Confidence Level
Numbers in Parentheses are Standard Errors

Source : World Resources Institute Calculations

"there is no reason to believe that energy subsidies promote economic growth." Similarly, no correlation was found between energy taxes and inflation. Energy consumption per unit of GNP a measure of energy inefficiency is, however, inversely correlated to energy prices, as shown in Figure 10.4. Goldemberg et al. (1987), notes:

> *Despite the distortions imposed by subsidies to conventional energy-supply industries, many consider eliminating them to be politically infeasible. Some who take this view propose similar subsidies for investments in energy efficiency. However, in most attempts to neutralize one market distortion with another, the compensating measure only partially offsets the effects of the original distortion and may have unintended, undesirable side effects.*

Baumol and Wolf (1981) showed that subsidizing a new energy source could actually increase the consumption of conventional energy forms, principally energy consumed for input to produce the new energy source.

Since energy efficiency can be considered a "new energy source" in that it reduces the consumption of conventional forms of energy, Goldemberg et al. (1977) continue:

> *Instead of introducing new subsidies to promote energy efficiency, a more appropriate policy would be to promote fairness in energy markets by eliminating existing subsidies for energy supplies. This politically challenging approach should be coupled with public policies that would promote the minimization of life-cycle costs in market decisions relating to energy. In general, energy subsidies should be restricted to solving problems the market cannot solve—such as the energy problems of the poor and research and development.*

To eliminate direct energy subsidies, energy prices should equal long-run marginal costs of energy supply. In the case of electricity, the price should equal the cost of providing an extra unit of energy. If this is $0.09/kWh, say, a user saving 1 kWh saves $0.09, while the electric company reduces its expenses also by $0.09 by not needing to provide that kWh. Marginal costs are significantly higher than average costs of electricity generation, particularly for developing countries with rapidly growing demand requiring the construction of new power plants. Many industrialized countries use electricity tariffs based on marginal costs, but keeping the average tariff equal to the average cost of providing electricity (C. Dubois, Electricite de France,

Figure 10.4
Commercial Energy Efficiency and Energy Prices, 1983

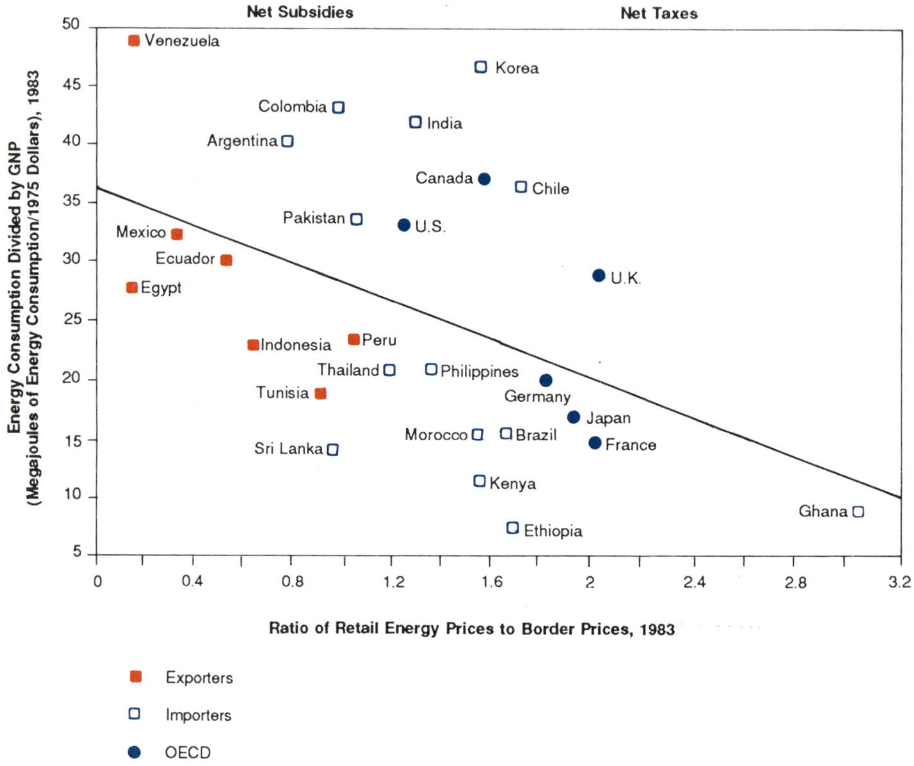

Fitted Regression Line	95% Confidence Interval for Slope of Line	Correlation Coefficient (x,y)
y = 37.09 - 8.18*x [All countries] (3.26)	-14.88, -1.48	-.45*

y = Energy Consumption GNP, 1983
x = Ratio of Retail Energy Prices to Border Prices, 1983
* = Statistically Significant at 5% Confidence Level
Numbers in Parentheses are Standard Errors

Source : World Resources Institute Calculations

pers. com., 1990). Thus some use categories pay less than the average cost and others pay more while keeping the entire utility operating without profit or loss.

Elevating the average price of electricity to the marginal cost would result in a surplus for the electric company, since marginal costs are significantly higher than average costs. The extra revenue can be used to finance investment in energy efficiency and new plant construction, and to deal with issues of social equity. For instance, connection charges may be reduced to permit connection of low-income households to the grid, provide a lower priced electricity for small consumers (a "lifeline" rate), or both without compromising marginal cost pricing. One plan calls for shifting a part of the residential subsidy to company fluorescent lamps, thus providing an incentive for low electricity consumers to use these lamps (Gadgil and Jannuzzi, 1989). For a discussion on long-run marginal cost pricing of electricity, see Sanghvi (1987) and Vedavali (1989).

Economically rational pricing should be extended to all energy sources to prevent uneconomic fuel switching, such as lighting with subsidized kerosene because electricity is highly priced.

Time-of-use rates. The marginal cost of providing electricity depends on the timing of the demand, since these costs are higher at periods of high demand. Tariffs that depend on the time of the demand provide an incentive to the user to save electricity during peak periods.Large users often have a separate meter that records the power demand (kW) and have to pay a separate "demand charge" per kW of maximum demand registered during the billing period. While this produces some incentive to reduce demand, it does not take into account the time of the demand, and whether it coincides with the utility peak.

A more rational tariff would use time-of-day and seasonal pricing to provide incentives that correspond to the costs of providing electricity at the different times. Developing countries are adopting time-of-day pricing for the largest industrial customers who represent most of the electric consumption. For smaller users, the expense of a complicated metering and billing system may not be warranted, and other arrangements are needed to regulate the timing of the demand. For instance irrigation pumping does not need to coincide with the daily peak demand; thus, electricity for irrigation may be provided during off-peak hours. Various technological options permit electricity companies to determine when power is being supplied to a remotely located device (e.g., by radio waves, power line ripples, etc.). These options may be adapted to the developing country situations.

Cogeneration policy. Industries with large demands for heat and power can simultaneously satisfy both through cogeneration. Using the waste heat on site, industrial cogenerators can often generate electricity at prices lower than the electric utility. To stimulate this production, electric utilities should be required to purchase cogenerated electricity at the utility's avoided cost, and should not penalize the cogenerator in other ways (e.g., by charging them a higher tariff to provide backup power). The enactment of such legislation a decade ago in the US led to enormous growth in cogenerated electricity, and offset the demand for new, capital-intensive central station power plants. Developing countries have advantages such as the possibility of cogenerating enormous quantities of electricity in sugar mills using advanced technologies (see Chapter 6).

Tax structure. Taxation policies may provide disincentives to the incorporation of energy-efficient products that are cost-effective from a social perspective. In Chapter 3, we noted that the extra cost of an energy-efficient air conditioner (without considering additional taxes) resulted in a cost of saved energy of $0.04/kWh; but, with a 100 per cent excise duty and a 25 per cent import duty on the extra cost of the more efficient unit, the cost of saved energy increased to $0.10/kWh, which is unlikely to be considered cost-effective. If the excise duty were the same total amount (not percentage), the disincentive towards energy efficiency would be removed, without the government losing revenue. (The question of import duties is discussed in the following section.) While an excise duty that is a fixed percentage of the sale price is easy to specify and administer, it is important to remove disincentives to energy efficiency. For motors we have proposed (Chapter 5) that the duty be based on input power of the motor, rather than sale price. The energy-efficient motor would have a lower input for the same output, giving it a slight positive incentive. Oversizing leads to energy waste, and this tax strategy would provide additional incentive to reduce oversizing. The level of the duty should be calculated on the basis of present sales by size in order to keep total revenue constant.

Import policy. Import duties should be considered within a broader framework of policies related to the importation of products and technical knowledge that have implications for future use. Over the last decade, India has opened its frontiers to the importation of products (automobiles, household appliances, office equipment, etc.) followed by a partial indigenization of their manufacture domestically. Giving priority to energy-efficient products makes domestic manufacturers more competitive economically and tends to improve their energy efficiency.

India's automobile industry, which enjoyed decades of economic protection from any form of foreign competition, provides an example. With the introduction of the Maruti, an 850-cc lightweight, Suzuki-design car with far lower fuel consumption than any domestically made car, other manufacturers soon took steps (through foreign collaborative arrangements) to improve the energy efficiency of their models. Technological advances in electronic household appliances (radio, television sets, videorecorders) have made these devices more energy-efficient over the years. The relatively late introduction of television sets and videorecorders into India has meant that the stock of these appliances is already relatively efficient. The Indian Standard IS 4547-1985 (black-and-white) and IS 10662-1983 (color), which limits the power consumption of television sets to 100 watts, have kept ostentatious models (projection and other large screen sets) off the market, although it is unlikely that these sets would have sold enough to make much difference.

For refrigerators, air conditioners, and washing machines, a wide range of efficiencies is available in the international market. Licenses for domestic manufacture of new models should take into account their energy consumption. One of the most effective strategies for achieving energy efficiency in these household devices is through energy-efficiency standards or agreements that limit maximum energy consumption. The import policy should be consistent and place a similar constraint on any device or technology being imported. As noted in Chapter 3, an error in selecting washing machine models now manufactured in India could have implications on future electricity demand. Such errors need to be avoided in the future.

Import duties should also consider the energy efficiency of the product and technology being imported. India already has a policy of import duty reductions for certain qualified energy efficiency (and renewable energy) products. This list could be periodically modified to take into account the importance of the end use, energy savings potential and cost-effectiveness. The subsidy implicit in the duty exemptions may be eliminated for those products where the benefits are small. A detailed analysis of import duty policies is beyond the scope of this review.

Financing Energy Conservation

Achieving an energy-efficient economy will greatly reduce requirements for capital needed if the business-as-usual scenario of power plant construction is pursued. Nevertheless, significant investments in energy efficiency will be needed. In this section, we consider various aspects of financing electricity conservation in a developing country.

Sources of capital. Facing capital shortages for the construction of power plants, many developing countries are considering private sector involvement in financing and operating power plants. Even in India, where only four privately-owned electric utilities remain today, the Government is encouraging increased private sector participation in electric power projects (BI, 1990). Clearly this vision needs to be expanded to increase private sector capitalization of electricity conservation projects as well. Indeed, because the costs to society of end-use efficiency are far lower than providing for increased demand through power plant construction, private capital should be more strongly invited to finance energy conservation projects.

Bilateral and multilateral international aid has been directed to the lower income developing countries for humanitarian reasons, for the development of export markets, and for geopolitical reasons. Recently, the global community has become convinced that human activities are changing the world's climate and, if they continue unchecked, could have serious consequences. The primary concern is global warming through the increased generation of carbon dioxide and other "greenhouse" gases. Energy conservation, particularly in developing countries where the energy efficiency is generally low, offers some of the lowest-cost options for reducing the emissions of carbon dioxide. Here, it is in the self-interest of industrialized countries to contribute capital to developing country energy conservation efforts. The recent EEC funding of India's Energy Management Centre (see Box 10.1) and the Program for Acceleration of Commercial Energy Research (PACER, see Box 10.2) funded by the US Agency for International Development are important steps in this direction.

International financial institutions play an important role in financing power projects in developing countries. In financing energy conservation, multilateral development banks and private banks can continue to be very important, by increasing the available capital and by providing foreign exchange that permits the purchase of critical elements and technologies in the international marketplace.

There is a difference, however, in the nature of financing. Whereas power plant loans are project-specific (i.e., financing the construction of a particular power plant or hydroelectric project, each project costing hundreds of millions of dollars), financing of energy efficiency will require distributing the investment over many smaller projects. Financing will need to be task-specific (with a significant amount of infrastructure development) rather than project-specific. Recent World Bank initiatives including an industrial energy conservation project for India (see Box 10.3) are major advances in this

Box 10.2
PACER: Program for Acceleration of Commercial Energy Research

PACER was set up in 1987 within the Industrial Credit and Investment Corporation of India (ICICI) for supporting technology innovation and development in India's energy sector. Apart from technology development in biomass, solar energy, and coal, PACER covers technologies to make better use of available generating capacity and to improve efficiency of major end-use electrical equipment in *industry, agriculture,* and the *commercial* sector.

PACER, funded by the US Agency for Industrial Development, promotes the *development of technology* through financial assistance to manufacturers, research institutions and end users. The main criteria for eligibility are the development of new or innovative products or processes with a significant potential for commercialization within five years. PACER assistance is limited to $3 million or 50 per cent of estimated cost. Payment to PACER is a negotiated percentage of revenues arising from project commercialization not exceeding 200% of the PACER contribution. PACER also provides grants for a *research* component of the development project as well as for relevant research projects.

(Source: PACER brochure, ICICI, Bombay, 1990).

Box 10.3
Industrial Energy Conservation in India – The World Bank Project

In recent years, the World Bank has made energy efficiency a priority of lending, with 29 energy efficiency initiatives, including eight stand-alone energy conservation projects. This would be the first such project in India, with $250 million in Bank financing with bilateral aid of $5 million each expected from Germany, the Netherlands, United Kingdom, and the United States, with other bilateral donors expressing a strong interest in supporting the program.

Program objectives

* Support and strengthen Energy Conservation Cell in the Department of Power and the Energy Management Centre (Box 10.1), to make them effective coordinating organizations.
* Develop and implement policies and programmes to improve incentives for efficient energy use, stronger institutions in public and private sector to carry out energy conservation programs, demonstration projects in key subsectors, and a program for research and development needs.

Project components

(1) (a) *Technical assistance* to Energy Management Centre and other agencies working in the field.

(b) Establishing an energy-sector database.

(c) Fund a number of studies in critical areas (taxation and subsidies, impediments to cogeneration, etc.)

(d) Promotional and informational schemes.

(2) *Demonstration projects* of industrial energy conservation, with the involvement of ICICI and IDBI

(3) Importation of *energy conservation devices*

Benefits

* Infrastructure building in improved industrial energy efficiency
* Direct benefits to projects supported including availability of needed imported devices during a period of severe balance of payments problems
* Boost demand for energy-efficiency investment
* Reduced degradation of the environment

Financial plan

Source	Amount ($ million)
World Bank	250
IDBI/ICICI	175
Bilateral donors	28
Government	51
Subloan recipients	125
Grant recipients	8.5
Total	637.5

(Source: *Energy Management News*, No. 3, June 1990).

direction. Different mechanisms for channeling the capital will be required, as well as different processes for evaluating that the funds are being well spent. These aspects of conservation financing are discussed below.

Channeling of capital. Since financing is not specific to individual large projects, it needs to be channeled through an agency that disburses the funds. This can be done through development banks, and, for electricity, through electric utilities.

India has a number of development banks which can finance energy conservation projects. These include Industrial Development Bank of India (IDBI), Industrial Credit and Investment Corporation of India (ICICI), Industrial Financial Corporation of India, National Bank for Agriculture and Rural Development (NABARD) at the national level, and industrial and rural development banks at the state level. IDBI (1990) has two schemes for financing energy conservation—one for Energy Audit Subsidy and one for Equipment Finance for Energy Conservation. While IDBI and ICICI and state-level equivalents can finance industrial conservation projects, NABARD could finance a major effort to improve the energy efficiency of irrigation pumps and rural industry.

There appears to be no current mechanism for financing electricity conservation in the residential and commercial sectors and city governments (street lighting, urban water and sewage pumping). The electricity distribution companies could be a mechanism for channeling financing (and for implementing) conservation projects among their smaller users. The utility has some natural advantages —they already know who the consumers are and how much energy they use, and they can recover their investment through their electricity billing system. Such a financing mechanism is specifically approved by The Indian Electricity (Supply) Act (1948) as amended by Acts 16 and 20 (1983):

> *Power of Board to engage in certain undertakings:*
> *(1) The Board may, in accordance with any regulations made in this behalf, manufacture, purchase, sell or let on hire on the execution of a hire-purchase agreement or otherwise, any electric machinery ... or apparatus for lighting, heating, cooling, or motive power or for any other purposes for which electricity can or may be used ... and may install, connect, repair, maintain or remove such ... apparatus ... and in respect thereof demand and take such remuneration or rents and charges and make such terms and conditions as it deems fit.*

> *(2) The Board may ... generally do all things including advertising, incidental to the sale and hire of such ... apparatus ...*
> *(3) The Board shall show separately in its accounts moneys received and expended by it in connection with any undertakings in which it engages under this section.*

After an initial period of resistance, many US electric utilities are actively involved in energy management projects (Clinton et al., 1986; Geller, 1989a; Goldman and Kahn, 1989).

Financial review of conservation projects. Since the products of energy conservation projects are not necessarily visible—an energy-efficient motor does not look very different from a conventional one—new mechanisms are needed for evaluating achievements and improving the confidence of the financial community in these projects. We have already noted the need for measuring electricity savings in conservation projects. This energy scorekeeping should be an element of financial review, which should also include actual projects costs, administrative and other project costs, loan recovery rates, etc. in determining the effectiveness of energy conservation projects.

Implementing energy conservation

With the information barrier, economic distortions, and capital availability attendant to energy conservation projects still need to be implemented. Here we present some implementation schemes that have proven particularly effective in other countries.

Equipment efficiency standards and protocols. Very high implicit discount rates in consumer decisions involving the extra costs of energy efficiency imply large losses to society in the long term. Household appliances is one example. Improved information to potential customers through efficiency labeling, in place since 1980 in the US, proved to be insufficient, and despite some resistance from the Reagan administration to "interfering with the market," minimum energy efficiency standards were legislated early in 1987. Japan, the EEC, and at least one developing country (Brazil) have implemented minimum efficiency standards for some household appliances. In Brazil, the government reached an agreement with the manufacturers rather than imposing a decree (Geller et al., 1990). By reducing adversarial relationships between government and industry, the protocol approach might be desirable for other developing countries as well.

Devices suited to standardization include incandescent and fluorescent lamps, lamp ballasts, refrigerators, air conditions, fans, other household appliances, electric motors, pumps, etc., where relatively few manufacturers dominate the market, and purchase decisions tend to be biased by the purchase cost. Standards or protocols are an inexpensive way of achieving energy efficiency. In India, standards exist for many of these devices. The strategy should be to make them universal rather than voluntary, and raise the standards to a level consistent with cost-effectiveness. (Testing should verify that the standards are indeed being met.)

Building standards. The energy required for air conditioning and space heating depends on the building characteristics and on the efficiency of the cooling or heating equipment. Since building designers, contractors, and landlords generally do not pay the energy bills, there is little incentive to offer energy-efficient designs to those responsible for paying energy bills. Thus, for new buildings, construction standards should take into account the expected energy use for space cooling or heating. One way to initiate building standards is to make them mandatory for government buildings and recommended for private buildings. This creates awareness of the standards, and market forces may provide incentives towards energy efficiency in private buildings as well. If market forces prove to be insufficient, standards may be made mandatory for all buildings. Prior to setting standards, of course, information needs to be available on the relationship between design and energy use, accompanied by a system for rating the energy efficiency of buildings.

In 1979, Singapore adopted efficiency standards for new commercial buildings that cover thermal transfer value for air-conditioned buildings, ventilation rates, indoor temperatures, and lighting power levels. Electricity use was reduced by 6 to 10 per cent (Chou and Ho, 1985).

Fostering energy management companies. Industries, existing commercial buildings, and possibly other energy user groups need information on how to save energy. Except for the largest industries, the task will be beyond the capabilities of existing staff. These companies and government agencies need specialized groups to provide the necessary services. At present, there are a number of government (e.g., National Productivity Council, Petroleum Conservation Research Association) and private energy management companies in India. More such companies will be needed for an expanded program on energy efficiency. Government could foster the creation of energy management companies by providing training, financing for the purchase of energy diagnostic equipment, and informational support services.

Direct installation. Some markets do not respond to energy efficiency standards or the available technical expertise. In some places, it may not be politically possible to reduce subsidies. In these instances, direct installation of the energy conservation measure by the electric utility is an alternative. In India, direct installation may be suitable to the rectification of irrigation pumps if removing the subsidy on electric rates proves difficult. The cost of direct installation to the electric utilities can be reduced if some part can be recovered through electric bills (see Loan programs below).

Hook-up fees. Any addition to the grid (such as a new office building) implies a capital inversion for the generation, transmission, and distribution of electricity to provide for the demand. A Brazilian strategy calls for charging an electric hook-up fee that depends on the building's energy use (Geller, 1990). A hook-up fee is an alternative to building efficiency standards, although it can be used with the standards to provide incentives for additional energy efficiency.

Energy guzzler taxes and efficiency incentives. An alternative or supplement to energy efficiency standards or protocols is taxes on the purchase of high energy-using devices or rebates on a low energy-using ones. While the standard sets a floor on the energy efficiency of an appliance, a graduated tax can give an incentive for achieving higher levels of efficiency. Sometimes the standard is set as an average efficiency level for all models sold by a given manufacturer, as in the case of US automobiles. A tax on products that are less efficient or a discount for those that are more efficient could produce an incentive to go beyond the average standard set. Many US electric utilities offer rebates for the purchase of energy efficient air conditioners, and other household appliances, principally to reduce peak demand (Geller, 1989b).

Loan programs. For many conservation projects, the availability of financing can improve the adaptation of energy conservation measures. We have already noted this in our discussion on channeling capital above. As long as electric prices to the end user are less than the marginal cost of supply, electric companies can afford to give some incentive (such as a lower-than-market interest loan) towards energy efficiency, and still reduce overall costs of supplying electricity. In India, the government may give the loan directly as a way of reducing its capital requirements for new power plant construction and related infrastructure equipments. IDBI's loans to industry for energy-efficient equipment purchase include an interest rate rebate below a nominal interest rate of 15 per cent as follows (IDBI, 1990):

> *A rebate of 2 per cent p.a. in the interest rate would be allowed if the savings in specific energy consumption due to the scheme is more than 10 per cent of the average specific energy consumption for the last three (3) years as certified by an approved energy auditor. If the energy saving is between 5 per cent and 10 per cent, a rebate of 1 per cent p.a. will be allowed.*

This is a good example of how the incentive is related to the (expected) energy savings. This reviewer would propose to measure actual energy savings, and modify the incentive scheme if necessary. Note that loans provided by the Indian Renewable Energy Development Authority do not include an incentive structure that depends on the extent of nonrenewable energy savings expected (or actually realized).

The IDBI loan provides a five-year period for loan repayment with a moratorium for the first two years. Longer repayment periods will provide positive cash flow for less cost-effective measures, and thereby permit larger energy savings.

At present, there are no loan programs for residential and commercial customers. One program for financing compact fluorescent lamps (CFLs) in households is the Bombay Efficient Lighting Large-Scale Experiment (Gadgil, 1990). Bombay Electric Supply and Transport Bombay's distribution utility will lease 36,000 CFLs to its customers for about $0.25 a month for four years. The lease payments repay the interest-free purchase cost of the lamps. The lamps save customers more in electricity than the lease payments. Since residential lighting is a significant contributor to peak demand in India (and other LDCs) and residential electricity is subsidized, Bombay Electric Supply and Transport will reduce costs by selling the "saved" electricity to higher-paying commercial and industrial customers. Maharashtra State Electricity Board will save by avoiding investments in additional generating capacity. A similar financing scheme for CFLs has been proposed for Karnataka (Reddy, Balachandra, D'Sa, 1990).

Appendix I

ECONOMIC GROUND RULES

An essential part of the end-use oriented approach is the use of economic analysis techniques that allow for a rational comparison among supply, end-use efficiency, and energy alternatives. In the present conventional approach, supply expansion, energy efficiency improvement, and renewable alternatives are planned and implemented by different government agencies using varied and inconsistent economic criteria. The budgets for these activities are determined independently, largely according to fashion or for political reasons. This, together with energy subsidies, market imperfections, and institutional obstacles, leads to a gross misallocation of resources.

The cost effectiveness of energy conservation is normally expressed in terms of a simple payback period (SPP, years) given by:

$$\text{SPP} = \frac{\text{Cost of conservation measure}}{(\text{annual energy savings}) \times (\text{unit price of energy})}$$

A measure is considered cost effective if SPP is sufficiently small. The numerator is the initial investment and the denominator is the annual energy cost savings to the user. Although this is a simple index, it has several drawbacks:

1. It does not consider the useful life of the conservation measure. A two year payback may be good for a measure that lasts 15 years but not for one that only lasts one year.

2. It does not consider the time value of money. The initial investment usually needs to be compared with savings realized over several years in the future, and a measure of the time value of money is essential.

3. The criterion is arbitrary—determination of the maximum payback period for a conservation measure to be considered cost effective is left to the user's discretion. There is no direct indication of how this investment in conservation might compare with alternative investment opportunities available to the user.

4. If more than one alternative is being considered, the smaller SPP appears to be a better investment. This is not necessarily so, as we shall see.

5. The price of energy includes taxes or subsidies, and thus does not necessarily equal the cost of providing the energy or of replacing the energy provided. Thus the SPP represents strictly the user's perspective and not that of the energy company or society at large.

6. Similarly, the energy price does not generally include the environmental or other costs of energy use, which are therefore left out of the decision making process.

Several alternative indices of cost effectiveness help overcome the first three drawbacks listed above. These include the internal rate of return (IRR), benefit to cost ratio, net present value, life-cycle cost, annualized life-cycle cost (ALC), and a concept unique to the end-use approach—the cost of conserved energy. All of these take into account the time value of money, in terms of a discount rate, d: one dollar now is worth (1 + d) dollars year from now. The discount rate can be expressed in terms of a nominal value or a real one, i.e. adjusted for inflation. As the inflation rate of a country varies, so does the nominal discount rate. The real discount rate tends to be more stable and reflects the real scarcity of capital. Unless otherwise stated we will use real discount rates throughout this report. We will also use fractions of one rather than percentages, i.e. a discount rate of 0.06 instead of 6 per cent.

The discount rate or the time value of money depends on the perspective. A small investor in a developed country has available investment alternatives that give a relatively low rate of interest; adjusted for inflation, the real return may be between 0 and 0.05. A conservation investment would be better if it gave a return only marginally higher. The "rational" discount rate of this investor would be the real interest rate available from the best avaialble investment alternative. A business investor has available higher return alternatives, and correspondingly has a higher "rational" discount rate. For someone who has to borrow money in order to pay for the conservation measure, the conservation is only cost effective if the return is higher than the interest rate on the loan—this in effect raises the "rational" discount rate. Generally, the discount rates for investment in less developed countries are higher than in industrialized countries, because of the shortage of capital.

A discount rate can be calculated from how people and institutions make investments among alternatives that have costs and benefits spread in time. This is called an "implicit" discount rate. For instance a consumer who favors a lower priced item even though its operating costs might be higher than a competing more expensive model has a higher discount rate than one who chooses in favor of lower operating costs. As noted above, a rational discount rate can be calculated for any perspective depending on alternative investment or loan possibilites available. Implicit discount rates are different from and usually higher than the corresponding "rational" values, for a number of reasons. The decision maker may be unaware of the true operating costs and tends to give it less importance than the well defined initial cost. For energy conservation investments, the implicit discount rates are often very high, because the energy costs of operating certain devices such as refrigerators are not well recognized, or because of real or perceived uncertainty about the energy savings actually to be realized in the "efficient" device.

It is now generally agreed that the rational discount rate should be the basis for policy making. The combination of factors that lead to implicit discount rates being higher than the rational values is usually referred to as market imperfections and call for remedial measures, be they informational or fiscal. In this report all economic comparisons will be based on the real rational discount rate. In later chapters we will discuss appropriate intervention strategies to deal with departures from this discount rate in actual investment decisions.

Before proceeding with the development of economic criteria for the evaluation of energy conservation measures, we first define the concept of the capital recovery factor (CRF):

$$\text{CRF} = \frac{d\,(1+d)^n}{(1+d)^n - 1}$$

Multiplying an investment that lasts n years by the CRF converts it into an annual expenditure equivalent, using the specified discount rated.

Internal Rate of Return

The IRR is the value of the discount rate at which two alternatives being considered have the same present value. In the case of a conservation investment between a conventional and an energy efficient alternative with investment costs of CC and CE, and annual energy use of EC and EE respectively, the IRR is the value of d obtained by solving the following equation:

$$CC + PE \times EC \sum_{1}^{n} \frac{1}{(1+d)^{k}} = CE + PE \times EE \sum_{1}^{n} \frac{1}{(1+d)^{k}}$$

where PE is the energy price and n is the life cycle of the investment. The series is the sum of the present value of energy costs incurred from 1 to n years in the future. This equation is nonlinear and it is usually solved in an iterative process. Different values of d are assumed and both sides of the above equation are calculated. An interpolation rapidly leads to a solution.

The advantage of this method is that a discount rate does not need to be specified since the result appears as an interest or return rate derived from the investment. One problem is the difficulty in comparing alternatives of different useful life, e.g., an incandescent lightbulb of 1000 hour life and a fluorescent that lasts ten times as long. This problem can be circumvented by further analytical complexity—comparing several purchases spaced over time of the shorter lived alternative to equal the life of the longer one. Another drawback is that only two alternatives can be compared at one time.

Cost of Saved Energy

The cost of saved energy (CSE) is a measure developed especially for energy conservation investments and permits the comparison of energy efficiency and energy supply alternatives on a rational basis. The cost effectiveness of the investment is expressed as the cost equivalent to a unit of energy saved. The CSE takes into account the discount rate and the useful life of the investment. It is calculated by dividing the annualized extra initial cost of the efficient alternative by the annual energy savings:

$$CSE = CRF(d,n) \quad (CE - CC) \;/\; (EC - EE)$$

In this case the price of energy need not be specified. The CSE calculated here may be compared with the price of energy—if the CSE is lower the conservation measure is cost effective from the perspective of the user. The CSE can also be compared with the cost of producing and distributing a unit of energy. This latter cost may include an amount to take into account an import premium where the energy itself or the equipment needed for its production and conversion needs to be imported. An environmental cost can also be added to the energy production cost. If the CSE of a particular conservation investment is lower than the (adjusted) unit cost of supplying energy, then from a societal perspective, the conservation investment is more cost effective. The societal perspective will be different from the individual user perspective in that energy prices may not reflect true costs (because of subsidies, taxes, and external costs), and because the individual and societal discount rates are different. Investment decisions in large energy projects are usually based on a smaller discount rate than that used by most users in their investment decisions.

The CSE also suffers from the same drawbacks as the internal rate of return (IRR)—alternatives with different useful lives are not directly comparable and more than two alternatives cannot be compared. Nevertheless, the cost of conservation energy is a powerful tool that makes possible a rational comparison among supply and conservation alternatives. Reddy has estimated the costs of producing and distributing a kWh of electricity using a variety of centralized and decentralized generation technologies and saving a kWh of electricity using several conservation options.

One advantage of the CSE approach is the ability of generating "supply curves." Here the horizontal axis represents the total potential of supplying or saving energy using a variety of supply or conservation options. The vertical axis is the cost of saved or supplied energy. The supply curve can be used to establish a least-cost energy strategy. Reddy has developed an electricity supply curve for Karnataka (Chap 7).

Annualized Life-cycle Cost

While the CSE is a powerful index that allows a rational comparison among a wide range of alternatives, this index has a number of drawbacks as we have noted earlier. For any end use, for instance irrigation pumping, only two alternatives may be readily compared. Also if there are several alternatives with different useful lifetimes, a rational comparison involves a good deal of analytical complexity. These problems are overcome by comparing the ALCs of the various alternatives. The ALC of each alternative is given by:

$$\text{ALC} = \text{CRF}\,(d,n)\;\; \text{C} + \text{PE} \times \text{E}$$

where PE is the energy price as before, C is the initial investment and E is the annual energy use of the alternative being considered.

In calculating IRR, the energy price but not the discount rate is specified, while for CSE, the discount rate but not the energy price is specified. The calculation of ALC, on the other hand, requires the specification of both energy price and discount rate. Since all costs are annualized, alternatives with different useful life can be readily compared, using an appropriate CRF for each case. Moreover, any number of mutually exclusive alternatives may be compared—each ALC is calculated, and the alternative with the lowest ALC is the most economical.

Where there are several mutually exclusive alternatives, SPP, IRR, and CSE require the comparison by pairs. This not only involves many different pairs—there are $n(n\text{-}1)/2$ combinations among n alternatives—but it is also misleading in that the alternative with the lowest SPP or CSE, or highest IRR is often not the most cost effective, i.e. it is not the alternative with the highest net present value or lowest life-cycle cost.

One way to avoid an incorrect conclusion would be to order the alternatives in increasing costs and energy savings (assuming they all have the same useful life), and the index calculated for each adjacent pair in succession.

Let (C1, E1), (C2, E2), (C3, E3), ... represent the investment costs and energy use, respectively, of a set of alternatives ordered such that C1, C2, C3, ... are in increasing order and E1, E2, E3, ... are in decreasing order. (Clearly, any alternative that has both higher cost and higher energy use than another can be eliminated from the list.)

Now the CSE (or the IRR) could be calculated comparing the second and first alternatives. Let this be designated CSE (1,2). If this were judged cost effective, then one would calculate CSE (2,3) to determine if the extra investment in option 3 is warranted by the extra savings. If so, one would continue with the next most expensive option, and so on. When the next extra step is no longer cost effective, say CSE $(m,m+1)$ is not cost effective, option m is likely to be the most cost effective. But this is not necessarily the case. Option $(m+2)$, at slight additional cost (compared to $m+1$ may promise much more energy savings). To determine this, CSE $(m,m+2)$ needs to be evaluated for cost effectiveness, and if found to be so $(m+2,m+3)$ needs to be compared and so on. A similar procedure may be adopted to determine the most cost effective use on the basis of IRR. For mutually exclusive alternatives corresponding to a given energy end use, the ALC approach is much simpler.

The ALC, on the other hand, cannot be used to compare different, independent end uses. We cannot directly determine if an energy-efficient lighting system is a better investment than an energy efficient refrigerator. Nor can conservation and supply alternatives be compared to obtain a global optimum as in the energy supply curves illustrated above.

The relative advantages and disadvantages of CSE and ALC imply that both need to be used, the first where conservation measures corresponding to different end uses need to be compared to one another and to energy supply alternatives, and the second where several mutually exclusive conservation measures (i.e., corresponding to the same end use) need to be compared to one another. The first is thus an important planning tool, while the second is more useful for detailed analyses of energy end uses.

Limitations of Economic Analysis

An analysis of cost effectiveness of an investment in energy efficiency does not take into account the non-financial aspects of the various alternatives. We have already mentioned some of these externalities —the requirement for foreign exchange and environmental consequences of energy projects. Large dams and other large-scale projects often require that populations be moved, causing economic hardship and cultural damage. The employment impacts may differ among alternatives. Qualitative differences might also exist among the alternatives.

Appendix II

INFORMATION SYSTEM IN URBAN PUBLIC TRANSIT: THE GREATER BUENOS AIRES EXPERIENCE

An excellent example of information in public transport is provided by Buenos Aires, an urban area of about 10 million inhabitants. Buenos Aires is also a diversified-integrated city. Its 200 neighborhoods include a mix of residential, commercial, and small industry and in some ways function more as individual villages or small towns than as parts of a single large urban area. Not only do people generally work near their homes, family members tend to live within the same neighborhood, thus reducing the needs for social travel as well.

The area is served by five subway lines, a light rail, a dense network of suburban railways, and a comprehensive bus system. Capital cost constrains most developing (or industrialized) countries from replicating the Curitiba system. Nevertheless, the information system is well worth disseminating. Urban street maps (available from several private publishers) show subway and rail lines and identify the buses that go between any two points in the city. Figure II.1 shows facing part of a guide (*Gui-Pla,* nd).

The map is divided into numbered grids which help readers to locate streets using a street index in the guide. The chart shows the bus, subway, and rail routes that pass through each grid number. By comparing the bus numbers serving an origin and a destination grid, one can readily identify a common bus serving both. For example, grid 13 is served by bus numbers by bus numbers 15, 28, 29, etc. Number 28 connects all grids on this section of the map. Similarly, route 151 connects grids 5 and 13. The map does not show the precise bus route within the grid. Elsewhere in the guide, however, is the complete route (Fig. II.2). Clearly, a single bus, rail or subway line does not connect all grid locations within the metropolitan area. However, Buenos Aires' routes are structured so that most origins and destinations require at most one connection: one first finds the routes connecting origin and destination to an appropriately located transport node. Bus stops are identified by bus number and indicate the entire route. Thus, even without a guide, one can usually find a bus that will stop near one's destination or a transportation node.

Because of Buenos Aires' excellent transport system, private cars are often not used even by people who do or could own them. There are virtually no expressways within the urban area. Nevertheless, there is little congestion, and even in the absence of any programs for automotive emissions control, the outdoor air quality is generally excellent.

Appendix 11.2

Facing Pages of A Transportation Guide (Gui Pla, Buenos Aires)

CUADRO NO.	TRANSPORTE QUE PASA POR EL PLANO 33 LINEAS NUMERO
13	15-28-29-42-130-151
14	15-28-29-42-130
15	15-28-29-57-60- 17-130-F.G.B.M.
16	15-28-29-57-60-117
17	28-33-37-42-107-160
18	28-37-42-160
19	
20	

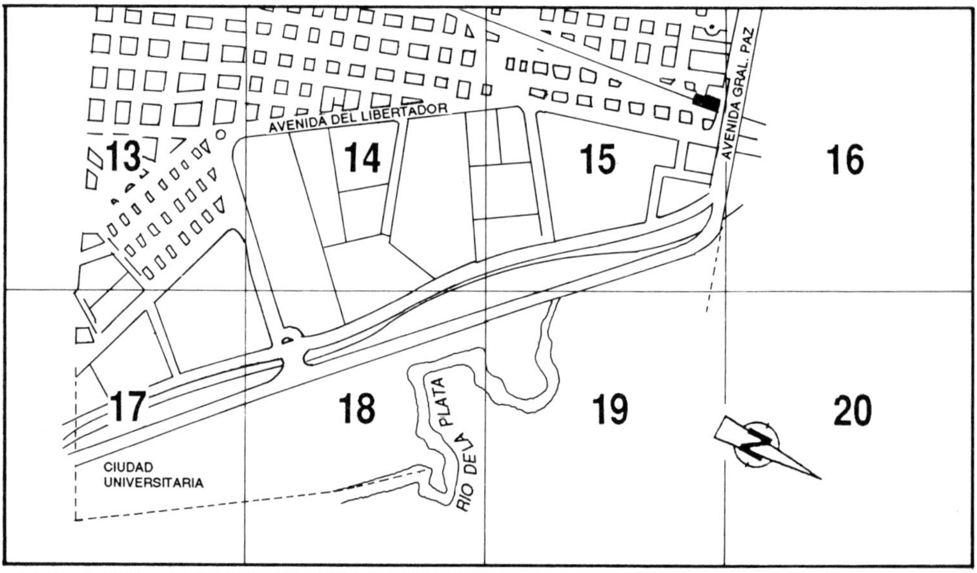

Appendix II.2
The Complete Route of the ISI Bus, Buenos Aires

Plaza Constitution – Est. Aristobulo del Valle

De Plaza Construction por:
Brazil – Lima 1700–1800 – Av. Juan de Garay 1100–1700 – Filberto 1700–1800 – Av. Entre Rios 1600–100 – PLAZA DEL CONGRESO – H. Yrigoyen 1800 – 1500 – L. Saenz Pena 100–1 – Ave3 Rivadavia 1500–3800 – PLAZA MISERERE al 2800 – Salguero 1–1200 – Ave Cordoba 3700–6300 – Jorge Newtaury 3500–5100 – Frado 400–1400 – Av Elbano 3100–0900 – Vitay del Prio – Vidal 1600–3000 – Cuesada 2800–2600 – Amenabar 3000–3400 – luna Azurduy 2600–2800 – Vidal 3400–4600 – Deheza 2700–2400 – Av. San Isidro 4700–4900 – Av. Cabildo 4900–500 – PUENTE SAAVEDRA – PARTIDO DE VICENTE LOPEZ – Av. Maipu – Est. A DEL VALLE – A. del Valle – 25 de Mayo – hasta Adolfo Alsina donde estacions.

De 25 de Mayo y Alsina (Partido de Vicente Lopez) por:
A. Alsina – Av. Maipu – Est. A DEL VALLE – PUENTE SAAVEDRA – CAPITAL FEDERAL – Av. Cabildo 5000–4800 – Pico 2400–2800 – Av. Cramer 4700–1700 – Av. de los Incas 3000–3300 – Conde 1600–500 – Maure 3200–3500 – Ave. Alvarez Thomas 400–1 – Av. Niceto Vega 6000–4600 – Av. Scalabrini Ortiz 1200–1300 – Jose A. Cabrera 4600–3800 – Medrano 1100–100 – Mitre 3900–1900 – PLAZA MISERERE AL 2800 – Riobamba 100–1 – PLAZA DEL CONGRESO – Corrbale de los Pozos – 1–1500 – Av. Pavon 1900–1200 – Lima 1500–1700 – hasta Brasil – PLAZA CONSTITUCION – donde estaciones.

Appendix III

A STRATEGY FOR RESOLVING INDIA'S OIL CRISIS

Amulya Kumar N. Reddy

(Reprinted from Current Science, January 20, 1981, Vol. 50, No. 2, 50-53)

[India is in the midst of an oil crisis. Many long-terms solutions have been suggested. The question that is being asked is: Can something be done immediately? Prof. A. K. N. Reddy, who leads the group on the Application of Science and Technology to Rural Areas at the Indian Institute of Science, has come up with a simple solution which appears to be well within our present technological capability. At our request Prof. Reddy has written a brief article for "Current Science" on this novel idea. The idea is so unorthodox and provocative that it merits discussion. We hope to publish comments of our readers on this article in a later issue.]

(Professor, Department of Inorganic and Physical Chemistry and Convener, ASTRA, Indian Institute of Science, Bangalore 560012, and Secretary, Karnataka State Council for Science and Technology, IISc, Bangalore 560 012, India)

India, like all oil-importing developing countries, is heading to a catastrophe due to the post-1973 escalations in oil prices. Unfortunately, most of the measures to tackle this crisis have been based on a narrow sectoral approach ignoring the interdependence of energy-consuming sectors.

This paper is an attempt to explore the inter-dependence of the transport and household sectors and to show that the resolution of the oil crisis lies in an alternative two-pronged strategy involving measures to be taken in the household sector.

India's oil consumption has been steadily increasing—the average annual compound rate of growth has been about 7 per cent since 1975–1976 (i.e., after the 1973 oil-price hike!). Indigenous production only accounts for about one-third of the country's total consumption which was about 33 million tonnes in 1980. Since two-thirds of the country's requirements are imported, the country's oil import bill has been escalating rapidly and has reached about 80%

of the estimated total foreign exchange earnings in 1980–1981. This already impossible situation can only worsen—hence, it is a matter of the country's survival to formulate and implement a strategy for the resolution of the oil crisis.

About 80% of the country's oil is used in the transport (60%) and household (20%) sectors. The transport sector uses motor gas (petrol), aviation transport fuel (ATF) and diesel—in 1978, these accounted for about 12%, 9%, and 79% of the consumption in this sector. Diesel is used by trucks/buses, railways and shipping, but trucks (64%) and buses (22%) consume about 86% of the total diesel in the transport sector.

Diesel locomotives haul about 37% of the country's freight whilst consuming about 11 per cent of the diesel used in the transport sector, in contrast to trucks which consume 64% of the diesel (6 times the railway freight consumption) to carry 36% of the freight. This only illustrates the well-known energy efficiency of railway transport compared to truck transport. Despite this, the share of the total freight carried by railways has been steadily decreasing—in 1960-1961, railways carried 84 per cent of the freight; today, the figure is about 64%.

Truck transport has been able to increase its share of freight mainly by increasing the average lead distance over which it carries freight. It is well known, however, that the costs of both truck and rail transport (in terms of resources utilized to move any commodity) increase almost linearly with distance, but truck costs, which are lower than rail costs at small distances, rise more rapidly with distance and exceed rail costs at large distances. In other words, there are break-even distances below which truck transport utilizes less resources, and above which rail transport is more economical in terms of resources. These break-even distances decrease as the diesel price increases --but, on the basis of a 50% increase over 1979 diesel prices, the break-even distances are between about 100-130 km depending upon the commodity. In fact, the average lead distances over which trucks are presently carrying freight are 3-4 times the break-even distances.

The main reason for this proliferation and expansion of truck transport—apart from factors such as shortage of wagons, delays, inefficiency, etc., which are not unavoidable features of railway transport—is the fact that financial costs to the truck operator are lower than the resource costs to society because of significant subsidies. An indication of the element of subsidy is the low price of diesel relative to petrol. In India, diesel has been only about a third to a half of the price of petrol, in contrasts to industrialized countries where diesel and petrol have almost the same price.

An apparently obvious device to shift freight traffic from truck to rail is to increase the price of diesel so that, for instance, it is on par with the price of petrol. But, this is not possible at present because the price of diesel is pegged at the price of kerosene which is also a middle distillate in refineries like diesel and substitutable for it in diesel engines. In other words, if diesel prices are increased without equal increases in kerosene prices, trucks shift to kerosene—as has been the country's experience in the recent past. Further, kerosene prices cannot be increased without causing great hardship to the poor because, at present, kerosene is used almost wholly in the household sector.

Out of this household consumption of kerosene about 70% goes for lighting and the rest for cooking. 87% of rural households and 38% of urban households, together constituting three-quarters of the country's 116 million households, depend wholly on kerosene for lighting. The amount of kerosene used is large—about 4 million tonnes in 1978-1979 in comparison with about 10 million tonnes of diesel. Further, about 40% per cent of the country's kerosene is imported. The consumption for lighting is, on the average, about 2.2 litres/month/household. Kerosene lamps have an extremely low luminous efficiency (output of light energy [in lumens] per unit of energy consumed in the lighting device)—the efficiency of electric bulbs (incandescent lamps) is about 200 times the efficiency of kerosene lamps.

Though electric lighting will result in a dramatic improvement in the quality of life, the situation with regard to electrification of homes is alarming. (a) All the 2,700 towns (with a population above 10,000), but only 44% (less than half) of the 567,000 villages have been electrified. (b) Only about 16% of India's electricity goes to rural areas. (c) Even this rural electricity is used primarily (to the extent of 87%) for agriculture (pumpsets) with only 13% being used by households -- hence, rural electrification has mainly served the pumpsets of the more affluent farmers. (d) The percentage of households electrified in electrified villages is only about 14%. (e) While the number of new electricity connections increases at the rate of about 1 million households per year, the number of new households increases at the rate of about 2.2 million households per year, i.e., the number of unelectrified homes is continuously increasing despite the decreasing number of unelectrified villages.

The first prong of an alternative strategy is to electrify all homes and provide them with electric lighting. Kerosene will then become unnecessary as an illuminant -- thus removing the constraint on diesel prices which can then be increased, and for instance made equal to petrol prices. If, in addition, the permit system for trucks and buses is designed and enforced to keep these vehicles within their break-even distances, there will be a drastic reduction in oil consumption. The National Transport Policy Committee has shown that, using realistic estimates of freight traffic, the projected diesel consumption for the year 2000 decreases from about 30 million tonnes to about 14 million tonnes by shifting to rail 75% of the inter-regional traffic moving by road beyond break-even distances (calculated on the basis of $30/barrel oil prices). But, oil prices may go up from $50 to $100/barrel in the next few years, implying that break-even distances will be even lower; and a much greater percentage, say 95% of inter-regional freight traffic allocable to rail can be shifted to rail—hence, much greater reductions in oil consumptions can be achieved. Thus, it seems possible to bring oil consumption to within the capability of indigenous production which is about 13 million tonnes/year.

Even the major reduction in oil consumption achievable by the first prong of the strategy may not be adequate because (a) intra-regional or short-haul traffic is best achieved by road transport and (b) for a sustainable future, the dependence of road transport must shift to renewable energy sources, i.e., to alternative fuels. Hence, the necessity of the second prong of the strategy.

Of the various renewable energy fuels, those that can be derived from biomass, viz., producer gas, methanol, biogas and ethanol—can easily be used in transportation vehicles. Ethanol, the Brazilian response to the oil problem—is produced from crops like sugarcane which require good agricultural land, leading to serious competition with food production—hence, ethanol is an unsatisfactory solution for a high population density country such as India. Biogas is a gaseous fuel which cannot be easily liquefied—it is only suitable for transportation over short distances from the biogas plant. The only feasible alternative fuels, therefore, are producer gas and/or methanol (which is obtained from producer gas). In fact, producer gas generated in situ with charcoal gasifiers was used in buses and trucks throughout the country during World War II.

But, both producer gas and methanol require wood resources, and the wood crisis in India is just as serious as the oil crisis. At present, about 130 million tonnes of firewood per year are consumed for cooking, and it is well known that firewood is becoming more and more difficult to produce. Hence, any attempt to use wood for producer gas and/or methanol in trucks will create even greater shortages of firewood and aggravate the problems of the poor for whom it is the main cooking fuel.

Firewood, however, is not an efficient cooking fuel—the efficiency of firewood stoves (chulas) is only about 5-10%. More efficient cooking fuels will not only provide greater convenience to housewives, but also conserve firewood. Fortunately, India has a large number of cattle—the country's human-cattle ratio is bout 2:3. The large amount of cattle waste can be used to generate biogas through anaerobic fermentation, and this biogas can be supplied to every rural home to meet its cooking needs (about 0.87 m3/household day). The cooking needs of urban homes, which are now being met by firewood, kerosene, soft coke and LPG (used in 73%, 17%, 5% and 5% of the homes respectively), can be met, for example, by piped mixtures of sewage, coal and producer gases. If gaseous cooking fuel is thus provided to all homes, not only will the 30% of the kerosene presently used for cooking become unnecessary, but the 130 million tonnes of firewood now being used for cooking will become available for other purposes, and in particular, as a renewable biomass fuel.

Out of this 130 million tonnes of firewood, about 75 million tonnes is sufficient to completely replace (after considering efficiency losses) the 10 million tonnes of diesel now used in trucks and buses.

Further, water-lifting pumpsets can be run on producer gas or methanol, and the 4.4 million diesel pumpsets projected for the year 2000 (present number = 2.7 million) would only consume about 16 million tonnes of wood to substitute completely for the 3.5 million tonnes of diesel which they would require. Also, if no further electrical pumpsets are connected beyond the present 3.6 million, and the anticipated increase of 7.4 additional pumpsets by the year 2000 are run on producer gas, about 27 million tonnes of wood are required. Thus, a total of 43 million tonnes of wood are sufficient to replace all the diesel that would be used in diesel pumpsets as well as the 17 TWh of electricity that would be required to run the 7.4 million extra pumpsets that are expected to be installed between now and 2000.

Hence, the 130 million tonnes of firewood released by providing sophisticated gaseous cooking fuels to all homes is more than the 118 million tonnes of wood required for all diesel trucks and buses as well as all the diesel pumpsets and additional electric pumpsets. In fact, once the pressure on trees as a source of cooking fuel decreases, the conditions for growing energy forests will be drastically improved leading to a much greater availability of firewood.

Finally, the 17 TWh of electricity saved by not increasing the number of electric pumpsets beyond the present number is only a little less than the 21 extra TWh which would be required to electrify all the unelectrified homes in the country. Thus, the net increase in electricity consumption by electrifying all homes and keeping electric pumpsets at their present number would be 4 TWh in 2000. It is not at all essential that the extra electricity for home electrification should come via grids from centralized generation—decentralized generation from local sources such as biogas gensets, microhydroelectric plants and energy forests can make a major contribution.

The two-pronged strategy described above shows that by the provision of electric lighting and efficient gaseous fuel to all homes, the country can move towards a virtually oil-free road transport system along with a dramatic improvement in the quality of life for its people. In fact, it appears that the country has been engulfed by a grave oil crisis because it has ignored two crucial basic needs of households: efficient energy sources for lighting and for cooking.

Appendix IV

REVISITING A STRATEGY FOR RESOLVING INDIA'S OIL CRISIS

Amulya Kumar N. Reddy

(Department of Management Studies, Indian Institute of Science, Bangalore 560 012)

An article with the above title was published about a decade ago in an Indian scientific journal, Current Science, and reprinted in the New Scientist of UK and the Bulletin of the Atomic Scientists of USA. The article did not attract any discussion in India, even from the energy and planning establishments, thought it offered hope in the gloomy aftermath of the second oil shock of 1979. Why? Perhaps because it was end-use-oriented, holistic and radical, in contrast to official energy decision-making in India which is supply-obsessed, sectoral and incremental. But, the more important reason was that the decision-making environment for energy changed drastically after the publication—hence, the tough decisions demanded by it seemed unnecessary.

It began to look in the 1980s as if the oil crisis had been warded off. But this was only an illusion created by (1) the spurt of crude oil production from Bombay High resulting from a deliberate Government decision, and (2) the fall in international oil prices due to production from non-OPEC sources. The situation also looked better than it was because of the bonanza of remittances from our workers in the Gulf. In fact, only the symptoms of the oil crisis had disappeared. The root causes remained, and were actually aggravated by crucial governmental decisions that ignored these causes.

Today, the oil crisis has again come to dominate the stage. Official measures (e.g., no sales between 7 pm and 7 am) and private responses (e.g., panic buying of petrol) are reminiscent of the 1970s after the oil-price hikes. Now, like then, the country is heading for an economic catastrophe due to the rise in oil prices.

But we must not be fooled by the feeling of deja vu into thinking that once again the crisis will go away. We are in a much graver situation than ten years ago. (1) In the past few years, output from Bombay High has almost peaked and indigenous production has been largely stagnant. Consequently, internal production is making a declining contribution to consumption. (2) We have squandered

Bombay High—and we have little to show for it—even though it belonged as much to future generations of our countrymen. (3) International oil prices have been on the rise recently mainly because non-OPEC oil production (which depressed oil prices in the early 1980s) levelled off a few years ago. (4) In the past few years, mopeds, scooters, motorbikes and cars have proliferated (thanks to deliberate Government policies) and their consumption of petroleum products can be reversed only with great difficulty. (5) The exodus of Indian workers from Kuwait and Iraq has led to a fall in remittances. This process may be repeated in the other countries of the Middle East. So, remittances may dry up. (6) Finally, we have wasted a valuable decade.

Except for the remittances, all the other reasons have nothing to do with the recent events in the Gulf. Thus, the Iraqi annexation of Kuwait is only the immediate cause of our current oil crisis.

The fundamental cause of our grave problem still is our unchecked appetite for oil. India's oil consumption has been rapidly increasing—the average annual compound rate of growth has been about 6% per year over the past 10 years. The 1989-1990 consumption of crude oil was 51.94 million tons (and estimated to rise to 58 million tons in 1990-1991) compared to 33 million tons in 1980. Indigenous production accounted for about two-thirds of the annual consumption of crude oil in 1989-1990 compared to only one-third a decade ago, but the 19.60 million tons of crude imported last year was 20% higher than a decade ago. So, it looks as if we used Bombay High only to meet the increase in our consumption—not to reduce our dependence on imported oil.

The country's import bill for oil and petroleum products was Rs 64,000 million in 1989-1990 constituting 23% of our total exports. However, every $1 per barrel increase in the international oil price is estimated to add an extra Rs 4000 million to our import bill and the price has already risen by $7.5 per barrel in August 1990. At this rate, our oil imports may, in the near future, account for over 60-70% of our exports—as was the case a decade ago.

There are two main reasons for the uncontrolled growth of our oil consumption.

- Diesel consumption is growing because of the increase in the total volume of freight traffic coupled with its shift from rail to road.
- Kerosene consumption is increasing because the number of non-electrified homes is increasing.

Railway Freight Deemphasized

The transport sector consumed about 48% of the total oil used in the country in 1987-1988. And, diesel accounts for 76% of the oil used in the transport sector, in comparison with petrol which is only 18%. The bulk of the diesel consumption in the transport sector is by trucks. In 1986-1987, trucks consumed 42% of the diesel. Trucks hauled 48% of the country's freight in comparison with 31% carried by diesel locomotives. But, to carry 1.6 times more freight, trucks consumed about 9 times more diesel than the railways (for freight). This only confirms the well-known energy efficiency of railway haulage of high-bulk density goods, compared to truck transport.

Despite this, the share of total freight transported by trucks, the more inefficient mode, has increased enormously since Independence, both in relative and absolute terms. In 1950-1951, trucks carried only 6% of the total freight of 105 billion ton kilometers; in 1989-1990, the percentage had risen to 52% of 433 billion ton kilometers.

Why are truck operators winning the competition for freight even though rail freight costs society less? The knee-jerk reaction is to blame the inefficiency of our railways—shortage of wagons, delays, pilferage of goods, inconvenience because of the lack of door-to-door service, unreliability, etc. But all these features of railway operations today are remediable.

The "success" of trucks is primarily because the price of diesel in India is subsidized and pegged at the price of kerosene. Thus far, diesel prices could not be increased without roughly equal increases in kerosene prices because, if the kerosene price was much lower than that of diesel, trucks adulterate their diesel fuel with kerosene and immediately create a kerosene shortage. This causes great hardship to the poor because kerosene is used almost wholly in the household sector, about one-third for lighting and the rest for cooking. And for the same reason, kerosene prices cannot be increased under present conditions.

Homes Not Electrified

In 1986-1987, about 100 million households (about 88% of rural households and 20% or urban households), i.e., two-thirds (63%) of the country's 142 million households, depended wholly on kerosene for lighting. Out of about 6.6 million tons of kerosene used in the country in 1986-1987 (in comparison with about 16.4 million tons of diesel) about 2.2 million tons was used for lighting. The average consumption for lighting is about 2.2 litres/month/household.

Kerosene lamps have an extremely low luminous efficiency -- even the efficiency of inefficient electric incandescent bulks is about 50 times greater than the efficiency of kerosene lamps. Electric lighting, therefore, will result in a dramatic improvement in the qualify of life.

Despite this, the number of unelectrified homes in the country is increasing at the rate of about a million households per year. This is because the total number of households is increasing at the rate of about 2.9 million households per year whereas the number of electrified households is only increasing at the rate of 2.1 million per year. Under these conditions, the country has been forced to increase kerosene consumption at the rate of 7.8% per year.

Petrol, A Small but Growing Problem

India's oil problem, therefore, is primarily a problem of the two middle distillates, diesel and kerosene, in that order. Thus, diesel and kerosene together account for as much as about half of our oil consumption, and incidentally account for the bulk of cut imports of petroleum products. In contrast, petrol represents less than one-tenth (actually 6.6%) of our oil consumption today (computed to 6.3% a decade ago). So, the popular belief that petrol-using vehicles—cars, scooters, etc.—are the major cause of our oil problem is mistaken. And, the governments' attempts to tackle the oil crisis through measures to reduce petrol consumption may be good populist stuff, but they are totally misdirected.

However, petrol which in 1980 accounted for only 12% of the oil used in our transport sector, has increased to 18% today. Thus, petrol may be a small problem today, but it is a rapidly growing problem which must therefore be addressed now.

Personal Vehicles Preferred

The growth rate (about 4%) of the urban population is double the growth rate (about 2%) of the population at large. The growing urban population generates an increasing need for transportation from homes to working places and back.

There are two ways in which this transportation service can be obtained:

- *public transportation* based on suburban trains (if feasible), buses, feeder transport mini-buses and three-wheeler autorickshaws, and
- *personal vehicles* consisting of mopeds, scooters, motorbikes and cars.

The decision-makers in our country have not only failed to provide the required outlays on public transportation but they have also encouraged the proliferation of mopeds, scooters, motorbikes and cars. In effect, therefore, the planners and Government have "chosen" personal vehicles as the preferred mode of intra-city passenger movement. Even worse, the lower the outlay on public transport, the more inadequate will public transport become, and the more will the public be compelled to go in for personal vehicles.

With the increasing emphasis on personal vehicles, the consumption of petrol has almost doubled from 1.5 million tons in 1980-1981 to 3.0 million tons in 1988-1989. This increase is reflected in the slight increase in the share of petrol in the consumption of oil products from 6.3 per cent in 1980-1981 to 6.6 per cent in 1988-1989.

So the increase in petrol consumption arising from the emphasis on personal vehicles (mopeds, scooters, motorbikes and cars) for passenger movement, rather than public transport, is a potentially major issue in the Indian oil scene.

Four-Pronged Strategy

A strategy for reducing, and even eliminating, the country's dependence on oil must be based on reducing the demand for diesel and kerosene. But the strategy must also reduce petrol consumption because that will, if unchecked, become a serious problem in the future.

A four-pronged strategy for resolving India's oil crisis is therefore suggested here. It consists of

- shifting freight traffic from road to rail;
- replacing oil with alternative non-oil fuels, particularly biomass-derived fuels;
- shifting passenger traffic from personal vehicles to public transportation; and
- implementing efficiency improvements in the use of petroleum products.

Improving the Efficiency of Using Petroleum Products

Efficiency improvements in the use of the main petroleum products can be achieved not only through long-term measures but also straightaway by better house-keeping with measures publicized by the Petroleum Conservation Research Association.

In the case of diesel, for instance, the long-term measures would involve improvement in the fuel efficiency of the truck fleet through better engines (e.g., direct injection diesel engines), reduction in chassis weight, better streamlining, greater payload capability by using multi-axle vehicles, better roads, etc. In addition, the fuel efficiency of diesel buses can be improved by having special lower-power bus engines to carry the smaller payloads of passengers compared to freight. Immediately, however, 10-15% reduction in diesel consumption can be achieved through training truck and bus drivers with regard to fuel-conserving driving habits. And of course, proper and timely maintenance yields significant fuel savings. The approach to improvements in the efficiency of petrol vehicles is similar to those recommended for diesel vehicles—fuel-efficient vehicles, economical driving habits, good maintenance and better roads.

Substantial diesel conservation can also be achieved in diesel pumpsets where as much as 30% savings are possible with friction-less foot-valves and plastic piping. Kerosene savings in cooking are possible through more fuel-efficient stoves, better cooking practices and the use of pressure cookers. These measures are also useful in reducing LPG consumption.

Shifting Passengers to Public Transportation

Apart from improvements in the fuel efficiency of petrol vehicles, the reduction of petrol consumption also requires a change in the model mix for passenger traffic away from personal vehicles to public transportation.

The essential overall measures include (1) massive investments on the infrastructure for public transportation (buses, passenger trains, national and feeder airlines and all the associated facilities for passenger convenience), (2) restriction of the supply of personal vehicles (mopeds, scooters, motorbikes and cars), and 3) an export orientation to the country's production of these personal vehicles.

But special measures are also necessary for intra-city passenger movement. These would include (1) major increases in the number of buses, and where possible, suburban trains, (2) optimization of routing of buses depending upon the passenger flows from homes to workplace and back, (3) dedication of key routes to buses, (4) banning of non-bus non-emergency traffic, particularly truck and personal-vehicle traffic, on these key bus routes between say 7 am to 9 pm, and (5) designation of special lanes for slow-moving bicycle traffic.

Shifting Freight from Road to Rail

Diesel consumption in the transport sector is linked to kerosene consumption in the domestic sector. Hence, the crux of the first prong of the proposed strategy is a massive program of home electrification. When all homes are electrified, kerosene becomes unnecessary as an illuminant. Along with separate measures discussed below for replacing kerosene as a cooking fuel in cities, its price can be brought on par with that of petrol.

Once this is done, the subsidy on diesel can be removed. Since diesel and petrol cost roughly the same to produce, they should have the same price. This is the case in many industrialized countries such as the US and UK. India too can raise the price of diesel so that it is on par with gasoline. The funds used for implicit subsidies on kerosene and diesel (which are of the order of about Rs 200,000 million per year) must be diverted to the improvement of the railways' freight operations.

The increase of diesel prices is a necessary, but not sufficient condition for decreasing truck freight. However, the price increase would create a favorable environment for further steps. These would include (1) increase of taxes on trucks so that they make a fairer contribution towards the maintenance of roads and related infrastructure, (2) cancellation of all long-haulage, or inter-state, truck permits except for special cases, (3) issuing of truck permits only for intra-state and adjacent-state transport, and (4) incentives for linking up trucks with the railways' freight operations to provide door-to-door freight movement.

If these measures are successful in shifting 90% of the long-distance truck traffic to the railways, the diesel consumption can be reduced by about 40% from the 36 million tons in the year 2000 just projected by the Planning Commission to about 22 million tons.

A further saving of 1.1 million tons of diesel can be achieved by diverting half the inter-city rail passenger traffic to buses and using the rail haulage capacity thus released for freight, where it is far more efficient than truck transport. Incidentally, the diesel price hike would raise the costs of bus travel, so it is essential to use part of the revenues from the price hike to prevent increases in bus fares.

The combination of these strategies (of shifting freight from trucks to rail and short-distance inter-city passenger traffic from diesel locomotives to buses) can reduce the diesel demand in the transport sector from about 36 million tons in the year 2000 to about 21 million tons which is only about 10% above the present consumption.

Shifting to Biomass-derived Fuels

Even with a major shift of freight from road to rail, the oil problem would only be alleviated: it would not be eliminated. Intra-regional or short-haul traffic would still require road transport and, therefore, a considerable amount of oil. And, for sustainable development, the dependence of road transport on non-renewable oil must be eliminated almost completely.

Hence, if an oil strategy is to be comprehensive and complete, it requires over the longer term the much more radical solution of shifting to alternative fuels for transportation.

Hydrogen produced by solar photovoltaics may well turn out to be the transport fuel of the future. But right now, only the liquid fuels — ethanol and methanol—are widely applicable alternative fuels for the transportation sector because they can be distributed through the nation-wide network already established for petrol and diesel. Both ethanol and methanol can be obtained from biomass sources: methanol from cellulosic materials such as fuelwood and agricultural wastes; ethanol from sugarcane or other feedstocks such as maize, etc. Ethanol can also be produced from cellulosic feedstocks, although at a higher investment cost compared to methanol.

A synergistic coupling between the transport sector and the agriculture sector is therefore possible whereby the output of the latter can be converted into a fuel for the former. "Fuel farms" can be established to supply the needs of transportation in the same way that rural farms produce food for urban demands.

There are, however, competing demands upon land and therefore the possibility of conflicts. The fuel-food conflict can be avoided by turning to non-agricultural land for cellulosic resources, particularly fuelwood, to produce methanol and/or ethanol. But the alcohol-from-fuelwood solution to the oil crisis can aggravate the domestic fuelwood problem particularly for the poor.

Cooking fuel for homes, however, is one of the basic energy needs, and the satisfaction of this need has to be an essential feature of an overall development-oriented energy strategy. Hence, the solution to the oil crisis must be compatible with the solution to the fuelwood problem.

Fuelwood, however, is used for cooking in traditional woodstoves at extremely low efficiencies of about 10%. If alternative high-efficiency fuels are provided for cooking, or the efficiencies of fuelwood stoves are radically improved, then the resulting drastic reductions in fuelwood consumption can set free a vast fuelwood resource base for the production of liquid fuels for the transport sector.

Either the biogas or high-efficiency fuelwood-stove options or a mix of them (with biogas in some places, perhaps coupled with a dairy development program to increase cattle-waste yields, and improved woodstoves in others) can be introduced in villages. Since about 110 million tons of fuelwood are currently being used in rural areas, there should be a saving of between 110 to 55 million tons of firewood, the two values corresponding to the two extreme cases of all-biogas and all (improved)-woodstoves, respectively.

In cities and towns, LPG consumption for cooking has been growing rapidly. In fact, the 9% growth rate in the 1970s and 23 per cent in the 1980s, would have been even higher if there had been balanced growth in the availability of indigenous LPG, cylinders, pressure regulators, bottling plants and distribution networks to keep pace with the demand. In 1989-1990, about 2.16 million tons of LPG was available to 16 million households. Since about 35% of the country's LPG was produced from natural gas, there is considerable scope for the expansion of LPG supplies because about 30% of the natural gas production in 1988-1989 was flared and India has large reserves of natural gas. By adopting the LPG option, perhaps supplemented with sewage gas, i.e., the biogas generated from the treatment of urban wastes and currently being wasted by flaring, the 23 million tons of fuelwood now being used by cities and towns would become redundant.

In all, therefore, the provision of high-efficiency cooking fuels and/or devices in rural and urban areas would make available 78 to 133 million tons of fuelwood (according to 1978 demand figures), provided that all the firewood being used today for cooking can still be collected. This saved fuelwood can be converted into methanol. If diesel fuel in trucks and buses is replaced with methanol, then the only diesel demand from the transport sector will come from the railways, i.e., 2.5 million tons.

Conclusion

Energy planning is meaningless unless one asks: "energy for whom?" and "energy for what?" India has been engulfed by a grave oil crisis because it has ignored two crucial basic needs of poor households: efficient energy sources for lighting and for cooking. The oil strategy proposed above shows that by the provision of electric lighting and efficient cooking fuels/devices to all homes, India can move towards a virtually oil-free road transport system. The lesson is simple: "Look after the people, and energy will look after itself!"

REFERENCES

ABE
1985 Towards a Perspective on Energy Demand and Supply in India in 2004/5, Report of the Advisory Board on Energy, Government of India.

ABB
1989 Flexible Power Generation Systems and their Planning, *ABB Review,* June.

Ahuja, D.R., V. Joshi, K.R. Smith and C. Venkataraman
1987 Thermal Performance and Emission Characteristics of Unvented Biomass-burning Cookstoves: A Proposed Standard Method for Evaluation, Biomass, 12:247-70.

Amerasekera, R.M. and B.P. Sepalage
1988 Sri Lankan Stoves—Past and Present, *Boiling Point,* No 15, Intermediate Technology Development Group, Rugby.

Anand, S. and V.S. Kothari
1990 Characterization of Electric Motors in Industry and Energy Conservation Potential in Indizill, Tata Energy Research Institute, New Delhi.

Andrews, C.J.
1989 Anticipating Air Conditioning's Impact on the World's Electricity Producers, *The Energy Journal,* July.

Baguant, J. and R. Panray
1991 Low-grade ethanol as a cooking fuel substitute for kerosene: a case study in Mauritius, in *Energy - Environment - Development* (Ed. R.K. Pachauri, L. Srivastava, and K. Ihukral), Proc. 12th. Int. Conf. of Int. Assoc. for Energy Economics, Jan 1990, New Delhi. Vikas Publishing House, New Delhi, pp. 929-937.

Baldwin, S.
1986 New Opportunities in Electric Motor Technology, *IEEE Technology and Society Magazine,* March, pp. 11-18.

1987 *Biomass Stoves: Engineering Design, Development, and Dissemination,* Princeton University Center for Energy and

Environmental Studies, Princeton University and Volunteers in Technical Assistance, Arlington, VA.

1988a Energy-Efficient Electric Motor Drive Systems: 1. High Performance Materials, Adjustable Speed Drives, and System Design, Princeton University Center for Energy and Environmental Studies.

1988b Energy-Efficient Motor Drive Systems: Motor and System Diagnostics, Princeton University Center for Energy and Environmental Studies Report 93, February.

1989 Energy-Efficient Electric Motor Drive Systems, Electricity; *Efficient End-Use and New Generating Technologies, and Their Planning Implications* (Johasson et al., eds.), Lund University Press, Sweden, pp. 21-58.

Baliga, B.N. and S. Dasappa
1989 Economics of Wood Gasifier Based on Power Generation, pp. 55-65 in *Proc. Seminar on Power Generation Through Renewable Sources,* Karnataka State Council for Science and Technology, Bangalore.

Baliga, B.N., S. Dasappa, U. Shrinivasa and H.S. Mukunda
1990 Gasifier based power generation: *Technology and Economics,* ASTRA, Indian Inst. of Science, Bangalore.

Baumol, W. and E. Wolff
1981 Subsidies to New Energy Sources: Do They Add to Energy Stocks? *Journal of Political Economy,* 89(5) October.

Benniger, J.R.
1986 *The Controls Revolution: Technological and Economic Origins of the Information Society,* Harvard University Press, Cambridge.

Berman, S. et al.
1989 Lighting Systems Research, FY 1988 Annual Report, Center for Building Science, Applied Science Division, Lawrence Berkeley Laboratory, University of California, Berkeley.

Bernardo, F.P. and G.U. Kilayko
1990 Promoting Rural Energy Technology: The Case of Gasifiers in the Philippines, *World Development,* 18:565-74.

Bhatia, R.
1990 Diffusion of Renewable Energy Technologies in Developing Countries: A Case Study of Biogas Engines in India, World Development, 18(4)575-590, April.

BI
1990 Power Privatization (editorial) and The Power That Will Be (cover article), *Business India,* Nov. 26 to Dec. 9.

BIS
1990 Energy Saving Standards: A Status Report on Indian Standards Influencing Energy Conservation, Bureau of Indian Standards, New Delhi.

1989 IS 12615:1989. *Induction Motors—Energy Efficient, Three-phase, Squirrel Case—Specification,* Bureau of Indian Standards, New Delhi.

Bleviss, D.L.
nd Improving the Efficiency of New Light Vehicles; Its Importance for Thailand and Methods to Achieve It, *Energy Efficiency Strategies for Thailand: The Needs and the Benefits* (D. Bleviss and V. Lide, eds.), Report of a Conference Held at Pattaya, Thailand, March 1988; International Institute for Energy Conservation, Washington, DC.

Bleviss, D.L. and P. Walzer
1990 Energy for Motor Vehicles, Scientific American Special Issue—Energy for Planet Earth, pp. 54-61, Sept.

Blomen, L.J.M.J.
1989 Fuel cells, in *Electricity: Efficient End-Use and New Generating Technologies, and Their Planning Implications (Johansson* et al., eds.), Lund University Press, Sweden.

Bohi, D.R.
1981 *Analyzing Demand Behavior,* Resources for the Future, Johns Hopkins Press, Baltimore, MD.

Boiling Point
1989 Special Edition on Briquetting, Intermediate Technology Development Group, Rugby.

Boyle, S.,
1990 Transport and Energy Policies—Only Connect, *Energy Policy,* Jan./Feb. 1990, pp. 34-41.

Bussmann, P.J.T.
1988 Woodstoves: Theory and Applications in Developing Countries, Ph.D. Thesis, Eindhoven University of Technology, Eindhoven University of Technology, Eindhoven, The Netherlands.

CAG
1988 Report of the Comptroller and Auditor General of India for the year ended March 31, 1987, Union Government (Scientific Departments), Chapter III, Department of Atomic Energy.

Centre for the Application of Science and Technology to Rural Areas (ASTRA)
1982 Rural Energy Consumption Patterns—A Field Study, *Biomass,* 2:255-80.

Charcoal Production Improvement for Rural Development in Thailand
1985 Forest Products Res. Div., Royal Forest Department, Ministry of Agriculture and Cooperatives, Bangkok, Thailand.

Chou, S.K. and J.C. Ho
1985 A National Strategy for Energy Management in Singapore, *Energy* 10(9)1017-22.

Clinton, J., H. Geller and E. Hirst
1986 Review of Government and Utility Energy Conservation Programs, *Annual Review of Energy,* 11:95-142.

CMIE
1990 Centre for Monitoring of the Indian Economy, Worli, Bombay.

Copersucar
1989 *Proalcoal: Fundamentos e Perspectivas,* Sao Paulo, Brazil.

Craig, P.P.
1976 Mass Transportation, Energy Conservation, and City Structure: the Need for New Goals, pp. 172-193 in *Future Strategies for Energy Development: A Question of Scale,* Proc. of a Conf. at Oak Ridge, Oct. 1976, Oak Ridge Associated Universities, Oak Ridge, TN.

Current Energy Scene in India
1990 Centre for Monitoring the Indian Economy, Bombay.

DAE
1985 Report of the Committee for Pricing of Nuclear Power, Department of Atomic Energy, Government of India, February.

Das, T.K., J. Roy and D. Chakraborty
1990 New Technological Options in Energy Sector Cost Burden on Rural Households, pp. 468-72 in *Integrated Renewable Energy for Rural Development,* Proc. of Annual Congress of the Solar Energy Society of India, Tata McGraw-Hill, New Delhi.

Dasappa, S., V. Reddy, H.S. Mukunda and U. Shrinivasa
1985 Experience with Gasifiers for 3.7 kW Engines, *Ambio,* 14:275-79.

Dayal, M.
1989 *Renewable Energy—Environment and Development,* Konark Publ., New Delhi.

De Montalembert, M.R. and J. Clement
1983 Fuelwood Supplies in the Developing Countries, Food and Agricultural Organization.

DeCicco, J., J. Cook, D. Bolze and J. Beyea
1990 *CO_2 Diet For a Greenhouse Planet: A Citizen's Guide For Slowing Global Warming,* National Audubon Society, New York.

Destor, G., M. Miloff and L. Haas
1979 Urban Form and Energy Conservation, Final Report CR 2440-7 prepared for the (Canadian) Ministry of State for Urban Affairs, Middleton Associates, Toronto, Ontario.

Dhawan, A.S., R.B. Mahajan, C.K. Ganure, V.V. Kulkarni and A.G. Gote
1990 Constraints in the functioning of biogas programme in Maharashtra State, *Energy Management,* July-Sept.

DOE
1989 Technical Support Document: Energy Conservation Standards for Consumer Products: Refrigerators and Furnaces Report, DOE/CE-2077, US Department of Energy, Washington, D.C.

Dunkerley, J., I. Hoch, C. Gadhok and K. Thukral
1987 Energy and Transport—the Indian Experience, *Pacific and Asian Journal of Energy,* 1(1)1-12.

Dunkerley, J., M. Macauley, M. Naimuddin and P.D. Agarwal
1990 Consumption of Fuelwood and other Household Cooking Fuels in Indian Cities, pp. 92-99 in *Energy Policy,* January/February.

Dutt, G.S. and M.F. Fels
1989 Keeping Score in Electricity Conservation Programs, pp. 353-388 in *Electricity Efficient End-Use and New Generation Technologies and Their Planning Implications,* (Johansson, et al., eds.), Lund University Press, Sweden.

Dutt, G.S., J. Navia and C. Sheinbaum
1989 Technologia Apropiada para Cocinar còn Lena, *Ciencias,* Depto, de Fisica, Facultad de Ciencias, Universidad Nacional Autonoma de Mexico, Mexico City, 15:43-47.

Eckholm, E.
1975 *The Other Energy Crisis: Firewood,* Worldwatch Institute Washington, D.C.

1976 *Losing Ground,* Worldwatch Institute, Washington, D.C. *Economic Survey 1989-90*

1990 Ministry of Finance, Government of India, New Delhi.

Editorial Opinion
1990 *Energy Management,* J. of the National Productivity Council, New Delhi, 14(2) 4 April-June.

EMC
1989 EMC to Train Energy Managers and Auditors, *Energy Management News,* 1:4, Nov.

1990 *Energy Management News,* 2, March.

EPRI
1986 Very Small (Approx. 500 KW) Low-cost Gas-fired Power Plants, EPRI AP-4827, Electric Power Research Institute, Palo Alto, California, USA, November.

ERG
1986 *Energy Research: Directions and Issues for Developing Countries,* IDRC-250e, Energy Research Group, International Development Research Centre, Ottawa.

Escay, J. R.
1990 *Summary Data Sheets of 1987 Power and Commercial Energy Statistics for 100 Developing Countries,* Industrial and Energy Department Working Paper, Energy Series Paper No. 23, World Bank, Washington D.C., March.

EUT
1985 Test Results on Kerosene and Other Stoves in Developing Countries, Report of Eindhoven University of Technology, World Bank Energy Department, Paper 27, Washington D.C.

FAO
1987 FAO Production Yearbook, Vol. 41, FAO Statistical Series No. 82.

Fitzgerald, K.B., D. Barnes and G. McGranahan
1989 *Interfuel Substitution and Changes in the Way Households Use Energy: the Case of Cooking and Lighting Behavior in Urban Java,* Industry and Energy Dept. Report, World Bank, Washington, D.C.

Forest Survey of India
1987 *The State of Forest Report,* Ministry of Environment and Forest, Dehradun.

Fulkerson, W., R.R. Judkins and M.K. Sanghvi
1990 Energy from Fossil Fuels, pp. 82-89 in *Scientific American Special Issue Energy for Planet Earth,* September.

Gadgil, A.J.
1990 The Belle Project: Bombay Efficient Lighting Large-Scale Experiment— Overview, (unpublished).

Gadgil A. and G. Jannuzzi
1989 *Conservation Potential of Compact Florescent Lamps in India and Brazil,* Lawrence Berkeley Laboratory Report LBL-27210, University of California, Berkeley, CA.

GEDA/GEB
nd Wind Farm at Okha—Completion Report, Gujarat Energy Development Agency and Gujarat Electricity Board.

Geller, H.S.
1985 Ethanol Fuel from Sugarcane in Brazil, Ann. Rev. Energy, 10:135-64.

1989a Implementing Electricity Conservation Programs: Progress Towards Least-Cost Energy Services Among U.S. Utilities, pp.741-63 in *Electricity: Efficient End-Use and New Generation Technologies and Their Planning Implications* (Johansson, et al., eds.), Lund University Press, Sweden.

1989b Promoting Electricity Conservation in Developing Countries, American Council for an Energy Efficient Economy, Washington, D.C., October.

1990a Electricity Conservation in Brazil: Status Report and Analysis, Electrobras (Brasilia), University of Sao Paulo (Sao Paulo), and American Council for and Energy-Efficient Economy (ACEEE, Washington, DC), Aug., available in ACEEE.

1990b Note on Sales of Standard and Energy-Saving Incandescent Lamps.

Geller, H.S., J. Buarrque de Hollanda, M.J. Marques
1990 Opportunities for Conserving Electricity Through Efficiency Regulations or Accords, presented at the Fifth National Energy Congress, Rio de Janeiro, October.

Geller, H.S. and G.S. Dutt
1983 Measuring Cooking Fuel Economy, in Wood Fuel Surveys, Report GCP/INT/365/SWE, FAO Programme for Local Community Development, Rome.

Geller, H.S., J. Goldemberg, J.R. Moreira, R. Hukai, C. Scarpinella and M. Ysonizawa
1988 Electricity Conservation in Brazil: Potential and Progress, *Energy,* 13(6)469-83.

Geltner, D.
1985 Transport and Energy in Developing Countries, *Energy Policy,* 13(4)340-344.

Generator Gas - the Swedish Experience from 1938-45
1979 Solar Energy Research Institute NTIS/SP-33-140, Golden, CO.

Gill, J.
1986 A Preliminary Investigation of Alcohol-Fueled Stoves, Report of the Energy Research Group, The Open University, Milton, Keynes.

1987 "Improved Stoves in Developing Countries—A Critique", *Energy Policy,* 15:135-44.

Gimenez, A.A.
1989 Transportation Energy Conservation Activities in the Philippines, in *Energy Efficiency Strategies for Thailand: The Needs and the Benefits* (D. Bleviss and V. Lide, eds.), Report of a Conference Held at Pattaya, Thailand, March 1988, International Institute for Energy Conservation, Washington, D.C.

Goldemberg, J., T.B., Johansson, A.K.N. Reddy and R.H. Williams
1988 *Energy for a Sustainable World,* Wiley-Eastern Ltd., New Delhi.

Goldemberg, J. and R.H. Williams
1985 The Economics of Energy Conservation in Developing Countries: A Case Study for the electrical Sector in Brazil, in *Energy Sources: Conservation and Renewables,* D. Hafmeister, H. Kelly and B. Levi, Eds, American Institute of Physics Conference Proc 135, American Institute of Physics, New York.

Goldman, C. and Kahn, E.
1989 Comparative Assessment of the Demand-Side Management Plans of Four New York Utilities, Energy 14(10)615-628.

The Guardian
1989 Based on UK Ministry of Transport statistics, Nov. 30.

Gui-Pia (Buenos Aires City Guide Book)
nd Ubicandose, S.A., Buenos Aires.

Hertz
1990 Hertz Electronic Devices, Calcutta, product literature.

Hochgreb, S.
1985 *Methane Production from Stillage,* COMGAS, Sao Paulo.

Hyman, E.L.
1986 The Economics of Improved Charcoal Stoves in Kenya, *Energy Policy,* April.

IDBI
1990 *Energy Conservation Schemes,* Industrial Development Bank of India, Bombay.

Illich, I.
1974 *Energy and Equity,* Brochure and Directory of Members, Indian Pump Manufacturers Association, Worli, Bombay.

IMPA
1990 Brochure and Directory of Members, Indian Pump Manufacturers Assn., Worli, Bombay.

IS
1985 *Specifications for Domestic Refrigerators* (mechanically operated), Indian Standard IS 1476-1979, first issued 1980, revised 1985.

ITDG
1984 *Noflie I and Noflie II: A Portable Wood/Briquette Stove,* Intermediate Technology Development Group, Rugby.

1985 *Low Energy Briquetting,* Project 508 Report, Intermediate Technology Development Group, Rugby, December.

Jash, T. and D.N. Ghosh
1990 Biogas Production from Slaughter House Waste, pp. 367-69 in *Proc. National Solar Energy Convention,* Calcutta, December.

Jain, B.C. and F. Vyarawalla
1985 Biomass Gasification—Its Relevance to India and Development Work at Jyoti, Inc. *Proc. Second Int. Producer Gas Conf.,* Bandung, Indonesia.

Jiang, Z. C. Zhang, N. Wu, et al.
1984 High Primary Air Biogas Range Burner, pp. 1043-52 in *Proc. 1984 Int. Gas Res. Conf., Government Institutes,* Inc., Rockville, MD.

Johansson, T.B., B. Bodlund and R.H. Williams (Editors)
1989 Electricity: *Efficient End-Use and New Generating Technologies, and Their Planning Implications,* Lund University Press, Sweden.

Jones, D.W.
1988 Some Simple Economics of Improved Cooktove Programs in Developing Countries, *Resources and Energy,* 10:247-64.

Joseph, S., K.L. Shrestha, E. Pelinck, J.G. Campbell and T.N. Bhattarai
1984 Introduction of Improved Stoves for Domestic Cooking in Nepal, Field Doc. 10, His Majesty's Government, United Nations Development Program Food and Agricultural Organization Community Forestry Development Project, Kathmandu, Nepal.

Joshi, V.
1988 Rural Energy Demand and Role of Improved *Chulha, Energy Policy Issues,* Tata Energy Research Institute, New Delhi, 4:23-35.

Joshi, V., C. Venkataraman, and D.R. Ahuja
1985 Performance of Cookstoves: Measurements of Thermal Efficiencies and Emissions in *Proc. 2nd World Congress on Engineering and the Environment,* New Delhi, November, Tata Energy Research Inst., New Delhi.

1987 Emission Factors in Biofuel Burning Metal Cookstoves, Tata Energy Research Institute, New Delhi.

Kishore, V.V.N.
1988 Cooking Energy Systems — A Comparative Study, in *Energy Policy Issues,* Tata Energy Research Institute, New Delhi, 4:37-45.

Kjellström, B.
1985 Biomass Gasifiers for Energy Supply to Agriculture and Small Industry, *Ambio,* 14:267-74.

Kosmo, M.
1987 *Money to Burn? The High Costs of Energy Subsidies,* World Resources Institute, Washington, D.C.

KPC
1982 Project Report of the Mallapur Mini-hydel Scheme, Karnataka Power Corporation, Bangalore.

KPC
1985-86 Annual Report, Karnataka Power Corporation, Bangalore.

Ladisch, M.R. and K. Dyck
1979 Dehydration of Ethanol: New Approach Gives Positive Energy Balance, *Science,* Vol. 205, No. 31, August 3.

Larson, E.D.
1990 Center for Energy and Environmental Studies, Princeton University, personal comm.

1991 A Developing-Country-Oriented Assessment of Technologies and Costs for Converting Biomass Feedstocks into Modern Energy Carriers, *Draft Report,* Princeton University Center for Energy and Environmental Studies, Princeton, N.J.

Larson, E.D. and L. Nilsson
1989 *A Technology Menu for the Efficient End-Use of Electricity: Vol. 1: Motor Drive Systems,* Dept. of Environmental and Energy System Studies, University of Lund, Sweden.

Larson, E.D., M.H. Ross and R.H. Williams
1986 Beyond the Age of materials, *Scientific American,* 254(6)24-31.

Larson, E.D., R.H. Williams, J.M. Ogden and M.G. Hylton
1989 Biomass-Fired Steam-Injected Gas Turbine Cogeneration for the Cane Sugar Industry, *Proc. 20th Congress of the International Society of Sugar Cane Technologists,* Sao Paulo, Brazil, Oct. 12-21.

Lashof, D.A. and D.A. Tirpak
1989 Policy Options for Stabilizing Global Climate, *Draft Report to US Congress,* Office of Policy, Planning, and Evaluation, Environmental Protection Agency, Washington, D.C.

Leach, G.
1989 Biomass Energy, Welfare and the CO2 Problem, *Global Warming and Climate Change: Perspective from Developing Countries,* Proc. Int. Conf. (S. Gupta and R.K. Pachauri, eds.) pp. 189-194.

Leach, G. and M. Gowan
1987 *Household Energy Handbook: An Interim Guide and Reference Manual,* Technical Paper 67, The World Bank, Washington, D.C.

Leach, G. and R. Mearns
1988 *Beyond the Woodfuel Crisis,* Earthscan, London.

Leitman, J.
1987 Residue Utilizing: A Recent Experience from Africa, *Boiling Point,* 12:11-14.

Lerner, J.
1989 The Curitaba Mass Transmit System, International Institute for Energy Conservation, Washington D.C.

LFMG
1990 Compressed Natural Gas—a Diesel Substitute in New Zealand, Liquid Fuels Management Group, New Zealand. Report summarized in *Energy Management,* April-June, 14(2)26-33.

Lokras, S.S., D.S.S. Babu, S. Bhogie, K.S. Jagadish and R. Kumar
1983 Development of an Improved Three-Pan Cookstove, pp. 13-17 in *Proc. of the ASTRA Seminar,* Indian Institute of Science, Bangalore.

Louvel, R.
1987 Briquetting of Vegetable Residues, *Boiling Point* Intermediate Technology Development Group, Rugby, 14:15-19.

Marques, M.J.
1990 PROCEL (Brazilian National Electric Conservation Program), letter to the President of the Brazilian Association of Electrical and Electronic Industry, 11 January.

Masera, O.
1990 Sustainable Energy Scenarios for Rural Mexico: An Integrated Evaluation Framework for Cooking Stoves, Masters Dissertation, Energy and Resources, University of California, Berkeley.

McGowan, T.
1989 Energy-Efficient Lighting, pp. 59-88 in *Electricity: Efficient End-Use and New Generating Technologies,* (Johansson et al., eds), Lund University Press, Sweden.

McKillop, A.
1986 Energy Sector Investment in LDCs: The Credibility Gap Widens, pp. 318-328 *Energy Policy,* August.

Mills, G.E. and E.E. Ecklund
1987 Alcohols as Components of Transportation Fuels, *Annual Review of Energy,* 12:47-80.

McNutt, B.D.
1989 Examining the U.S. Experience with Fuel Economy Regulation and its Application to Developing Countries, Office of Policy Planning and Analysis, US Dept. of Energy, Washington, D.C.

Meier, P. and M. Munasinghe
1987 Implementing a Practical Fuelwood Conservation Policy: The Case of Sri Lanka, *Energy Policy,* April.

Mills, G.A. and E.E. Ecklund
1987 Alcohols as components of transportation fuels, Annual Review of Energy, Vol. 12, pp. 47-80.

Mintzer, C.M.
1987 *A Matter of Degrees: The Potential for Controlling the Greenhouse Effect,* World Resources Institute, Washington, D.C.

Mongia, N., R. Bhatia, J. Sathaye and P.B. Mongia
1990 Least-Cost Strategies for Reducing CO2 Emissions in India: Some Preliminary Results, Paper Presented at the Second International Workshop on Energy and Global Climate Change, Lawrence Berkeley Laboratory, Berkeley, California, Oct.

Moore, E.A. and G. Smith
1990 *Capital Expenditures for Electric Power in Developing Countries in the 1990s,* Industry and Energy Department Working Paper, Energy Series Paper No. 21, World Bank, February.

Moreira, J.R. and J. Goldemberg
1981 Alcohol—Its Use, Energy and Economics—a Brazilian Outlook, *Resource Management and Conservation* 1:231-279.

Mukerjee, S.
1988 Automobile Pollution Control Strategies, *Energy Environment Monitor,* Tata Energy Research Institute, New Delhi, 4(2)11-29, September.

Mukunda, H.S., U. Shrinivasa and S. Dasappa
1988 Portable Single-Pan Wood Stoves in High Efficiency for Domestic Use, *Sadhana,* Indian Academy of Sciences Proc. in *Engineering Sciences,* 13(part 4)230-70.

Munasinghe, M.
1983 Third World Energy Policies: Demand Management and Conservation, *Energy Policy,* 11(1)4-18, March.

Muthusubramanian, P.
1990 Presentation at the National Solar Energy Convention, Calcutta, December.

Muthusubramanian, P. and T.M. Haridasan
1990 Performance Evaluation of Biomass Gasifier Systems and Their Cost Effectiveness, pp. 351-354 in *Integrated Renewable Energy for Rural Development,* Proc. National Solar Energy Convention, Calcutta, December.

NAEE
1989 *The Netherlands Travelling Clean: Towards a Trend Breach in Passenger Transport,* Netherlands Agency for Energy and the Environment.

Natarajan, I.
1985 Domestic Fuel Survey with Special Reference to Kerosene, National Council for Applied Economics Research, New Delhi.

Nikura, T.
1986 Energy Efficiency in Japan, presented at the TERI workshop on Energy Policy Issues.

Norgard, J.S.
1989 Low Electricity Appliances—Options for the Future, pp. 125-172, in *Electricity: Efficient End-Use and New Generating Technologies,* (Johansson et al., eds.), Lund University Press, Sweden.

Ogden, J.M. and R.H. Williams
1989 *Solar Hydrogen: Moving Beyond Fossil Fuels,* World Resources Institute, Washington, D.C.

Ogden, J., Williams, R. H. and Fulmer, M.
1990 Cogeneration Applications of Biomass Gasifier/Gas Turbine Technologies in the Cane Sugar and Alcohol Industries, *Proc. Program for Acceleration of Commercial Energy Research* (PACER) Conf., New Delhi, April; revised version, June, Princeton University Center for Energy and Environmental Studies, Princeton, NJ.

Openshaw, K. and C. Feinstein
1989 Fuelwood Stumpage: Considerations for Developing Country Energy Planning, Industry and Energy Dept., Working Paper Energy Series Paper 16, World Bank, Washington, D.C.

Pachauri, R.K. and V. Suri
1990 Contribution to Greenhouse Gases Through Large-Scale Use of Fossil Fuels and Biofuels, *Energy Environment Monitor,* 6(1)11-16, citing D.R. Ahuja, Environmental Protection Agency, Washington, D.C.

Pai, B.R.
1989 Power Generation and Cogeneration Using Al-20 Gas Turbine Engines, *Proc. Seminar on Power Generation from Renewable Sources,* Karnataka State Council for Science and Technology, Bangalore.

Pandey, M.R.
1984 Domestic Smoke Pollution and Chronic Bronchitis in a Rural Community in the Hill Region of Nepal, *Thorax,* 39:337-39.

Pandey, M.R., J.S.M. Boleij, K.R. Smith and E.M. Wafla
1989 Indoor Air Pollution and Acute Respiratory Infection in Children, *The Lancet,* February 25; summarized in *Boiling Point,* (18)15-18, Intermediate Technology Development Group, Rugby.

Pandey, M.R., H.N. Regmi, R.P. Neupane, et al.
1985 Domestic Smoke Pollution and Respiratory Function in Rural Nepal, *Tokai J. Exper.* and Clin. Med., 10:471-81.

Paramesvar, K.R.
1988 Household Appliances—Energy Efficiency, Quality, and Standards, *Energy Policy Issues,* Vol. 4, Tata Energy Research Institute, New Delhi.

Patel, S.M.
1988 Low-Cost and Quick-Yielding Measures for Energy Conservation in Agricultural Pumps, *Pacific and Asian Journal of Energy,* 2(1)3-11.

Pearse, J.
1991 Technology choice, Chapter IV of draft report on Regional Technical Assistance on Environmental considerations in energy development, Asian Development Bank, Manila.

Philips, M.
1991 Lending for Energy Efficiency: Opportunities for the Multilateral Development Banks, draft report, International Institute for Energy Conservation, Washington, D.C.

Phiu-Nual, K.
1989 The Potential for Energy Savings in Thailand's Transportation Sector, pp. 275-87 in *Energy Efficiency Strategies for Thailand:* The Needs and the Benefits (D. Bleviss and V. Lide, eds.), Report of a Conference Held at Pattaya, Thailand, March 1988, International Institute for Energy Conservation, Washington, D.C.

Prabhakara Rao, K.V.
1989 The Potential for Cogeneration of Electricity in the Sugar Industry of Karnataka, Master's Dissertation Project, Indian Institute of Science, January.

Proalcool: *Fundamentos e Perspectivas,* 2nd Edition
1989 Copersucar, Sao Paulo.

Qiu, D., G. Shuhua, B. Liange and G. Wang
1990 Diffusion and Innovation in the Chinese Biogas Program, *World Development* 18:555-63.

Rahman, M.
1990 Washing Machines: a Boom is in the Offing, p. 139 in *India Today,* Nov. 15.

Raju, S.P.
1953 *Smokeless Kitchens for the Millions,* Christian Literature Society, Madras, India.

Rajvanshi, A.K.
1984 Distillation of Ethyl Alcohol from Fermented Sweet Sorghum Solution by Solar Energy, Final Project Report to Dept. of Non-Convention Energy Sources (DNES), New Delhi.

1988 Solarly Distilled Ethanol from Sweet Sorghum as Cooking Fuel, pp. 383-93 in *Food-Energy Nexus and Ecosystem,* (T.K. Moulik, ed.), Oxford and IBH Publ., New Delhi.

Ramakrishna, J.
1988 Patterns of Domestic Air Pollution in Rural India, Ph.D. Dissertation, University of Hawaii. Resource Systems Institute, East-West Center, Honolulu HI.

Ramesh, S. and T.V. Natarajan
1987 Policy Options in Rural Electrification in India, with Special Reference to Pumpset Energization, *Pacific and Asian Journal of Energy,* 1(1)45-53.

Ravindranath, N.H. and H.N. Chanakya
1986 Biomass-based Energy System for a South Indian Village, *Biomass,* 9:215-33.

Ravindranath, N.H., Nayak, M.M., Hiriyur, R.S. and Dinesh C.R.
1991 The Status and Use of Tree Biomass in a Semi-Arid Village Ecosystem, *Biomass* (in press).

Ravindranath, N.H., R. Shailaja and A. Revankar
1989 Dissemination and Evaluation of Fuel Efficient and Smokeless ASTRA stove in Karnataka, *Energy Environment Monitor,* 5(2)48-60.

Reddy, A.K.N.
1985a A Strategy for Resolving India's Oil Crisis, *Current Science* (India), 50(2)50-53.

1985b An End-Use Methodology for Development-Oriented Energy Planning in Developing Countries, with India as a Case Study, Princeton University Center for Energy and Environmental Studies Report 181.

1987 On the Loss and Degradation of Tropical Forests, paper presented at the International Conference on Tropical Forestry, Bellagio, Italy, July; Department of Management Studies, Indian Institute of Science, Bangalore.

1990 A Strategy for Resolving India's Oil Crisis, Dept. of Management Science, Indian Institute of Science, Bangalore.

Reddy, A.K.N., P. Balachandra, and A. D'Sa
1990 A Programme for the Replacement of Incandescent Bulbs with Company Fluorescent Lamps in the Domestic Electric Connections of Karnataka, Draft Report, Dept. of Management Studies, Indian Institute of Science, Bangalore.

Reddy, A.K.N., and J. Goldemberg
1990 Energy for the Developing World, pp. 63-72. Scientific American *Special Issue Energy for Planet Earth,* September.

Reddy, A.K.N. and B.S. Reddy
1983 Energy in a Stratified Society: Case Study of Firewood in Bangalore, *Economic and Political Weekly,* New Delhi, October 8.

Reddy, A.K.N., H.I. Somasekhar, P. Rajabapaiah and S. Jayakumar
nd Scenarios for the Evolution of the Pura Energy Centre, Report of ASTRA, Indian Institute of Science, Bangalore.

Reddy, A.K.N., G.D. Sumithra, P. Balachandra and A. d'Sa
1990 *Economic and Political Weekly* (India), June 2, pp. 1201-1216.

Reddy, A.K.N., Sumithra, G. D., Balachandra, P., and d'Sa, A.
1991 The Comparative Costs of Electricity Conservation and Centralized and Decentralized Electricity Generation, in *Energy - Environment - Development* (R.K. Pachauri, L. Srivastava, and K. Thukral eds.) Proc. 12th. Int. Conf. of Int.

Assoc. for Energy Economics, Jan. 1990, New Delhi. Vikas Publishing House, New Delhi, pp. 772-814, Table 6.

Renner, M.
1988 Rethinking the Role of the Automobile, Worldwatch Paper No. 84, Worldwatch Institute, Washington D.C., June.

RMI
1989 *Competitek: State of the Art:* Drivepower, Rocky Mountain Institute, Snowmass, Colorado, April 1989 Edition.

Rosillo-Calle, F. and D.O. Hall
1987 Brazilian Alcohol: Food Versus Fuel? *Biomass,* 12:97-128.

Salaria, P.N.
1987 Efficient Use of an Induction Motor, *Electrical India,* March 15.

Sangen, E. and P. Visser
1986 A Study of the Performance of Charcoal Stoves, *Boiling Point,* Intermediate Technology Development Group, Rugby, 10:15-18.

Sanghvi, A.P.
1987 Towards a More Efficient Framework for Rationalizing Power Tariffs, *Pacific and Asian Journal of Energy,* 1(2)47-58.

SERI
1979 Generator Gas—The Swedish Experience from 1939-45. Solar Energy Research Inst., Golden, CO, NTIS/SP-33-140.

Shukla, K.C. and J.R. Hurley
1983 *Development of an Efficient, Low NOx Domestic Gas Range Cook Top,* Report by Thermo Electron Corp., Waltnam, Mass. for Gas Research Institute.

Siemons, R.V., Ahmed and H. Hood
1989 Cotton Stalk Charcoal Agglomeration in the Sudan, *Boiling Point,* Special Edition on Briquetting, Intermediate Technology Development Group, Rugby.

Singer, H.
1961 Improvements of Fuelwood Cooking Stoves and Economy in Fuelwood Consumption, Report to the Food and Agricultural Organization, Rome.

Singh, R. and C.S. Prasanna Kumar
1989 Electronic Fan Regulator for Energy Conservation, *Energy Management,* New Delhi, 13(3)33-34.

Siwatibau, S.
1981 *Rural Energy in Fiji: A Survey of Domestic Rural Energy Use and Potential,* International Development Research Centre, IDRC-157e, Ottawa, Canada.

Smith, K.R.
1987a *Biofuels, Air Pollution and Health:* A Global Review, Prenum Press, New York.

1987b Cookstove Smoke and Health, *Boiling Point,* Intermediate Technology Development Group, Rugby, 13:2-9.

Sperling, D.
1989 *New Transportation Fuels: A Strategic Approach to Technological Change,* University of California Press.

Srinivasan, J.
1985 Construction and Performance of an Experimental Solar Pond Report to the Karnataka State Council for Science and Technology, March.

Srinivasan, M.R.
1985 Economic Case for Small and Medium Reactors in India, presentation to International Atomic Energy Agency.

Srinivasan, M.R. and K.V. Mahadeva
1987 The Comparative Investment and Financing Requirements for Nuclear and Coal-Based Energy Systems, Paper IAEA-CN-48/26, International Conf. on Performance and Safety, Vienna, Sept.-Oct.

Srivastava, U.K. and C. Sharan
1987 *Wind Energy for Water Pumping,* Oxford IBH Publishing Co., New Delhi.

Statistical Abstract of the United States
1989 109th Ed. Bureau of Census, US Dept of Commerce, US Govt. Printing Office, Washington D.C.

Statistical Pocket Book
1989 Central Statistical Organization, Dept. of Statistics, Ministry of Planning, Government of India.

Stix, G.
1990 Meteoric Messages: Keeping Tabs on Trucks, Volcanoes and the Nile, *Scientific American,* Sept., p. 119.

Sudhakar Reddy, B.
1990 PhD Dissertation, Management Studies, Indian Institute of Science, Bangalore.

Taylor, L.
1990 Lessons from Japan: Separating Economic Growth from Energy Demand, Association for the Conservation of Energy, London.

TERI
1987 Stove Efficiencies and Harmful Emissions, Report of Tata Energy Research Institute, New Delhi.

1989a Energy Management in Indian Railways, Tata Energy Research Institute report.

1989b *TERI Energy Data Directory and Yearbook,* Tata Energy Research Institute, New Delhi.

1990a Energy Efficiency of Domestic Appliances: Interim Report of Tata Energy Research Institute, to ICICI, Jan., New Delhi.

1990b presentation at the National Solar Energy Convention, Calcutta, Dec.

Tolpadi, S.G.
nd Energy Conservation in Agricultural Pumping Systems: Sights and Insights, Kirloskar Brothers Ltd.

Tulyanon, A.
1989 Energy Efficiencies in the Use of Compressed Natural Gas for Bangkok Buses, pp. 311-29 in *Energy Efficiency Strategies for Thailand: The Needs and the Benefits* (Bleviss and Lide, eds.), Report of a Conference Held at Pattaya, Thailand, March 1988; International Institute for Energy Conservation, Washington D.C.

Utria, B.E.
1989 Evaluacion del Sector Energetico Residencial en la Republica Dominicana, *Seminario Latinoamericano de Planificacion Energetica para el Sector Residential,* San Jose, Costra Rica; available from Industry and Energy Dept., ESMAP/World Bank, Washington, D.C.

Utilization of Sludge Gas for Operation of Gas Turbine Engines
1981 *Newsletter of the National Aeronautical Laboratory,* 9(11-12), Bangalore, India, Nov.-Dec.

van der Plas, R.
1988 Domestic Lighting. Working Paper WPS 68, Industry and Energy Department, World Bank, Washington D.C.

van Til, J.
1980 A New Type of City for an Energy-Short World, The Futurist, June, pp. 64-70.

Vasudevan, P. and S. Jain
1989 Field Level Study on Biogas Burners, *Changing Villages,* J. of the Consortium on Rural Technology, New Delhi, 8(2).

Vedavali, R.
1989 Domestic Energy Pricing Policies, Industry and Energy Department, Working Paper 13, World Bank, Washington, DC.

Vidyarthi, A. and V. Vidyarthi
1985 A Study on the Prospects of Using Liquid Fuels for Cooking in Developing Countries, Intermediate Technology Development Group, London.

Volunteers in Technical Assistance
1982 *Testing the Efficiency of Woodburning Cookstoves:* Provisional International Standards, Arlington, V.A.

von Hippel, F. and B.G. Levi
1983 Automobile Fuel Efficiency: the Opportunity and the Weakness of Existing Market Incentives, Resources and Conservation, 10:103-124.

Walsh, M.P.
1988 Pollution on Wheels, Report to the American Lung Association, February.

Williams, R.H.
1986 A Major Role for Developing Countries in Promoting Super Fuel-Efficient Cars?, presented at the OLDE/UNDP-DICD Workshop on Rational Use of Energy and End-Use Oriented Energy Strategy, Sao Paulo, Nov. 1985; revised version, Nov. 1986, Center for Energy and Environmental Studies, Princeton University, Princeton, N.J.

1988 Are Runaway Capital Costs a Constraint on Development? A Demand Analysis of the Power Sector Capital Crisis in Developing Countries, presented at the *International Seminar on the New Era in the World Economy,* The Fernand Braudel Institute of World Economics, Sao Paulo, Brazil, Aug. 31-Sept. 2, 1988. Cited edition: revised draft, Sept. 9, 1988, Princeton University Center for Energy and Enrironmental Studies.

Williams, R.H. and E.D. Larson
1989 Expanding Roles for Gas Turbines in Power Generation, pp. 503-53 in *Electricity: Efficient End-Use and New Generating Technologies* (Johansson et al., eds.), Lund University Press, Sweden.

WHO
1984 *1983 Statistics Annuals,* World Health Organization, Geneva.

Young, P.
1988 Promising Performance from a New Briquette Burner, *Boiling Point,* Special Edition on Briquetting, Intermediate Technology Development Group, Rugby, 16:8-9.

Zhu, H., M.R. Brambley and R.P. Morgan
1983 Household Energy Consumption in the People's Republic of China, *Energy,* 8(10)763-74.